Mapping Partition

RGS-IBG Book Series

For further information about the series and a full list of published and forthcoming titles please visit www.rgsbookseries.com

Published

Decolonising Geography? Disciplinary Histories and the End of the British Empire in Africa, 1948-1998
Ruth Craggs, Hannah Neate

Rescaling Urban Poverty: Homelessness, State Restructuring and City Politics in Japan
Mahito Hayashi

The Urban Question in Africa: Uneven Geographies of Transition
Padraig R. Carmody, James T. Murphy, Richard Grant, Francis Y. Owusu

Theory and Explanation in Geography
Henry Wai-chung Yeung

How Cities Learn: Tracing Bus Rapid Transit in South Africa
Astrid Wood

Defensible Space on the Move: Mobilisation in English Housing Policy and Practice
Loretta Lees and Elanor Warwick

Geomorphology and the Carbon Cycle
Martin Evans

The Unsettling Outdoors: Environmental Estrangement in Everyday Life
Russell Hitchings

Respatialising Finance: Power, Politics and Offshore Renminbi Market Making in London
Sarah Hall

Bodies, Affects, Politics: The Clash of Bodily Regimes
Steve Pile

Home SOS: Gender, Violence, and Survival in Crisis Ordinary Cambodia
Katherine Brickell

Geographies of Anticolonialism: Political Networks Across and Beyond South India, c. 1900-1930
Andrew Davies

Geopolitics and the Event: Rethinking Britain's Iraq War through Art
Alan Ingram

On Shifting Foundations: State Rescaling, Policy Experimentation And Economic Restructuring In Post-1949 China
Kean Fan Lim

Global Asian City: Migration, Desire and the Politics of Encounter in 21st Century Seoul
Francis L. Collins

Transnational Geographies Of The Heart: Intimate Subjectivities In A Globalizing City
Katie Walsh

Cryptic Concrete: A Subterranean Journey Into Cold War Germany
Ian Klinke

Work-Life Advantage: Sustaining Regional Learning and Innovation
Al James

Pathological Lives: Disease, Space and Biopolitics
Steve Hinchliffe, Nick Bingham, John Allen and Simon Carter

Smoking Geographies: Space, Place and Tobacco
Ross Barnett, Graham Moon, Jamie Pearce, Lee Thompson and Liz Twigg

Rehearsing the State: The Political Practices of the Tibetan Government-in-Exile
Fiona McConnell

Nothing Personal? Geographies of Governing and Activism in the British Asylum System
Nick Gill

Articulations of Capital: Global Production Networks and Regional Transformations
John Pickles and Adrian Smith, with Robert Begg, Milan Buček, Poli Roukova and Rudolf Pastor

Metropolitan Preoccupations: The Spatial Politics of Squatting in Berlin
Alexander Vasudevan

Everyday Peace? Politics, Citizenship and Muslim Lives in India
Philippa Williams

Assembling Export Markets: The Making and Unmaking of Global Food Connections in West Africa
Stefan Ouma

Africa's Information Revolution: Technical Regimes and Production Networks in South Africa and Tanzania
James T. Murphy and Padraig Carmody

Origination: The Geographies of Brands and Branding
Andy Pike

In the Nature of Landscape: Cultural Geography on the Norfolk Broads
David Matless

Geopolitics and Expertise: Knowledge and Authority in European Diplomacy
Merje Kuus

Everyday Moral Economies: Food, Politics and Scale in Cuba
Marisa Wilson

Material Politics: Disputes Along the Pipeline
Andrew Barry

Fashioning Globalisation: New Zealand Design, Working Women and the Cultural Economy
Maureen Molloy and Wendy Larner

Working Lives - Gender, Migration and Employment in Britain, 1945-2007
Linda McDowell

Mapping Partition

Politics, Territory and the End of Empire in India and Pakistan

Hannah Fitzpatrick

WILEY

This edition first published 2024

Registered Offices
John Wiley & Sons, Inc., 111 River Street, Hoboken, NJ 07030, USA
John Wiley & Sons Ltd, The Atrium, Southern Gate, Chichester, West Sussex, PO19 8SQ, UK

For details of our global editorial offices, customer services, and more information about Wiley products visit us at www.wiley.com.

Wiley also publishes its books in a variety of electronic formats and by print-on-demand. Some content that appears in standard print versions of this book may not be available in other formats.

Library of Congress Cataloging-in-Publication Data is Applied for:

Hardback ISBN: 9781119673804
Paperback ISBN: 9781119673835

Cover Design: Wiley
Cover Image: © Wikimedia Commons

Set in 10/12pt PlantinStd by Straive, Pondicherry, India

SKY10069978_031924

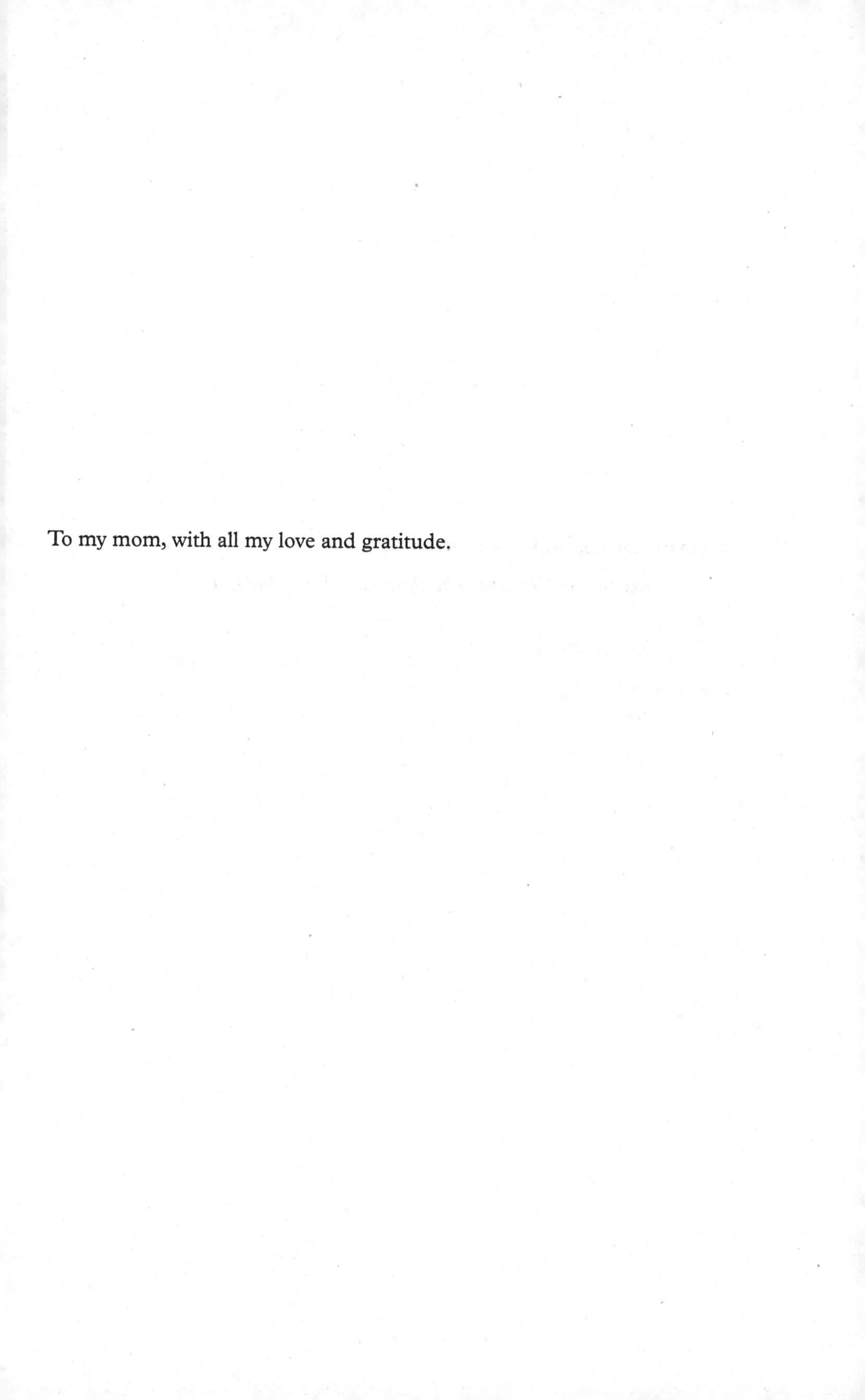

To my mom, with all my love and gratitude.

What the map cuts up, the story cuts across.

Michel de Certeau, *The Practice of Everyday Life*

Contents

List of Figures viii
Acknowledgements ix

1. **Remapping Partition** 1
2. **Surveying and Boundary-Making in Colonial India** 34
3. **Territorialising India and Pakistan** 77
4. **Geographies of the Punjab Boundary Commission** 114
5. **Oskar Spate, the Muslim League and Geographical Expertise** 139
6. **Partition to Partitions: New Avenues for Historical Geography** 181

Index 191

List of Figures

Figure 1.1	Partition Boundaries in the Punjab	3
Figure 2.1	Historical Maps of India, 1765 and 1805	52
Figure 2.2	Historical Maps of India, 1837 and 1857	53
Figure 2.3	Prevailing Religions of the British Indian Empire	59
Figure 2.4	Prevailing Races of the British Indian Empire	60
Figure 3.1	Continent of Dinia and its Dependencies	106
Figure 4.1	Map of the Punjab, 1931	123
Figure 4.2	Official Map of the Punjab Before Partition	126
Figure 4.3	The Congress Red Map	130
Figure 4.4	The Muslim League Map	131
Figure 5.1	Professor O.h.k. Spate	142
Figure 5.2	Draft Map of the Punjab from Spate's Collection	151
Figure 5.3	Spate's Sketch Map	157
Figure 5.4	Spate's Draft Map Depicting Three Possible Corridors	162
Figure 5.5	Map of the North Western Railway System	163
Figure 5.6	Spate's Draft Strategic Map	166
Figure 5.7	Draft Map by Spate of Sheikhupura Tehsil	172

Acknowledgements

Many people had a hand in supporting this book, as well as its previous iterations as postgraduate and undergraduate research. The initial idea for the project was developed during my time as an undergraduate at Barnard College in New York. It would never have seen the light of day without the support and training I received there, most especially from Rachel McDermott, who I want to be when I grow up. Anupama Rao taught me how to think about empire, while Jack Hawley taught me how to think about religion. Lisa Tiersten and Robert McCaughey nurtured my love for history. At St Andrews, I was supervised by the incomparable Dan Clayton, who believed in this project from the very beginning and who shaped my geographical education more than anyone else. I could not have had a better mentor and teacher.

The School of Geography and Sustainable Development, St Andrews, provided a small grant to obtain some of the archival materials reproduced here. A research incentive grant from the Carnegie Trust for the Universities of Scotland in 2016 provided funding for a research trip to the National Library of Australia. I extend my sincere thanks to the many archivists at the National Library of Australia and British Library, Chris Fleet at the National Library of Scotland and Kevin Greenbank at the University of Cambridge Centre of South Asian Studies. Virginia Spate and Andy Spate, with the kind assistance of Jane Mills, very generously granted permission for me to use materials from the Oskar Spate Papers, held at the National Library of Australia. It was a privilege to speak with Virginia Spate about her father's work and his memories of his time working for the Muslim League, and I am grateful to Andy Spate and the family for their continued support and assistance.

Thank you to the editors of the RGS-IBG book series, especially Ruth Craggs, who supported the manuscript (and me) throughout the process, and Dave Featherstone, who guided me through the proposal process. I am immensely grateful for their assistance and editorial guidance. Many thanks as well to the kind folks at the RGS, including Catherine Souch, Phil Emerson and Anna Lawrence.

Many geographers have supported this book in large and small ways. Satish Kumar and Mike Kesby were constructive and supportive examiners of the PhD, and Satish helped with the initial transition from PhD to book. Thanks also to James Sidaway, Tariq Jazeel, Andy Davies, Charlie Withers, Robert Mayhew, Miles Ogborn and Mike Heffernan. Stephen Legg has been a trusted reader and regular supporter, and I am grateful for his help over the years.

Thanks to colleagues in St Andrews, especially Matt Sothern and Sharon Leahy. I will be forever grateful to Lorna Philip and Caitlin Cottrill, who helped make time for me to visit Canberra while I was a teaching fellow at the University of Aberdeen. Caitlin was a mentor and advocate at a pivotal time in my life, and I will miss her very much. At the University of Edinburgh, thanks especially to Tim Cresswell, Fraser MacDonald, Andy Dugmore, Hayden Lorimer, Sukanya Krishnamurthy, Vaishak Belle, Faten Adam and Kath Will. Thanks also to the excellent historians who have answered questions and shared their knowledge, especially William Gould, Lucy Chester, Sumathi Ramaswamy, Gyanesh Kudaisya and Matthew Edney.

Anindya Raychaudhuri has been a steadfast colleague and friend since the moment we met, and both the book and my life would be diminished without his immense generosity. Thanks to excellent friends Alex Gnanapragasam, Angela Roberts, Hamish Kallin, Clare Raychaudhuri, Miriam Gay-Antaki and Emmy Pierce. My heartfelt thanks to Fiona, who created the extra space I needed.

My family has been rooting for me from the finish line for many years, including my late grandfather Jack, who remembered 1947 and believed in the importance of my work. Love and thanks to Dad, Terri, Mike and Kelli. Tom has been my most loyal champion and a true partner in every sense of the word. He asked, 'What about the dog?' and he is correct, Ketchup the cocker spaniel was my writing companion. She is the best good girl.

My final thanks are to my mom, who gave me everything. This book is for her.

Chapter One
Remapping Partition

Introduction

On 14 August 1947, Pakistan celebrated its first day of independence in its new capital, Karachi. Crowds of people celebrated in the street, raising and waving flags and cheering 'Pakistan Zindabad!' ('Long Live Pakistan!') as the new prime minister of Pakistan, Muhammad Ali Jinnah, drove through the streets in a State procession accompanied by the last British viceroy of India, Lord Louis Mountbatten. 'Spirits were high', one observer recalled, as 'officials who had opted for Pakistan were pouring in', already beginning the work of establishing the government and 'working in improvised offices' (Spate, 1991, pp. 58–59).

Despite the celebrations, the festivities 'were on a modest scale', charged with a sense of uncertainty, as the geopolitical partition of the subcontinent had only just begun to take shape (Spate, 1991, p. 59). The official map of the new borders had not yet been made public. People checked newspapers and public notice boards for graphic evidence of independence – for a map – and for official news to reach towns, villages and local government offices. And the uncertainty was accompanied by threats and outbreaks of violence: Mountbatten had reported news of a bomb plot targeting Jinnah during their official drive to Government House in Karachi on 14 August. Despite the threat, the new leader of Pakistan insisted on the procession, and so even as 'the route was lined with enthusiastic crowds', the police and military presence was a reminder of the very real fear and

Mapping Partition: Politics, Territory and the End of Empire in India and Pakistan, First Edition. Hannah Fitzpatrick.

uncertainty facing the new states of Pakistan and India (Mansergh 1983, p. 771). There were rumours of riots, explosions and scores of refugees crossing the new eastern and western borders of India and Pakistan. The celebrations were joyful, yet they were intermingled with fear, sadness and the imminent threat of violence.

Such fears were not unfounded. It has been estimated that roughly 14.5 million people migrated across borders in the four years following Partition, and that between two and three million people were killed or went missing (Bharadwaj et al., 2009). Kudaisya (1996) puts the number of migrants even higher, at over 18 million. The majority of migrants crossed the western boundary, with approximately 73% arriving in the Punjab (Waseem, 1999, p. 203). Those that arrived on both sides recounted stories of terrible violence, the abduction of women, and the killing, torture and rape of neighbours, relatives and strangers. Even for those who did not directly experience or witness the violence, the atmosphere was permeated by anxiety and insecurity. Even at the elite levels, the business of constitution-writing and nation-building that lay ahead was intertwined with a sense of territorial indeterminacy and uncertainty. The new borders were eventually announced on 17 August, two days after Indian independence and three days after Pakistani independence. On Independence Day, the border lines had been drawn, but only a select few, stationed elsewhere, knew their course (Figure 1.1).

This is a book about those borders. It is about how those borders were drawn and enacted in a rush in the summer of 1947, but also how they were imagined in multiple and contested ways during the prior decades of the 20th century, and about how their foundations were slowly and subtly brought into being in the 19th century. The book recovers and examines cartographic and territorial imaginaries of India at work from the late 19th century in order to trace the historically constructed discourses of ethnic, religious and geographical difference which permeated political discourses of distinctiveness and separation. Rather than follow the line of thinking summed up by Sumathi Ramaswamy that, 'although the Partition of India was arguably precipitated by nongeographic and extraterritorial factors, the actual determination and drawing of the new boundary lines was a cartographic act', I argue that geographical knowledge and colonial cartography were in fact at play prior to the actual drawing of the borders in July and August 1947 (2017, p. 290). In doing so, I construct a detailed story that takes a longer view of the geographical genealogy of Partition in order to trace the ways in which the geographical logic that was mobilised for the consolidation and administration of British India was the same logic that gave rise to the social, political and spatial conditions of Partition. The book also shows how a longer historical geographical account, which incorporates both the 19th and 20th centuries, can inform new ways of conceptualising the geographies of Partition, as well as uncover as-yet under-examined histories of geography in British India. This empirical investigation opens up new ways of understanding the events of 1947 as being more than simply a succession of interrelated decisions, negotiations

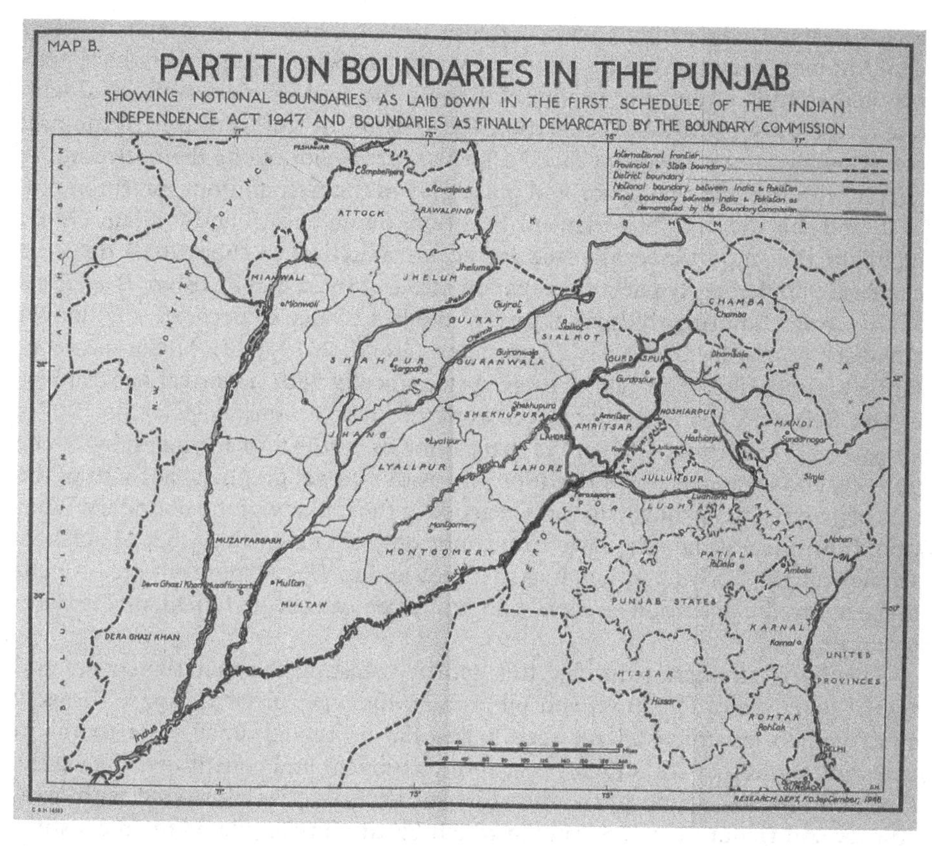

Figure 1.1 Partition Boundaries in the Punjab Map depicting notional boundary (east) and the boundary demarcated by Punjab Boundary Commission (west).

Source: British Library, Report on the last Viceroyalty, 22nd March-15th August 1947 by Rear Admiral the Earl Mountbatten of Burma, KG, PC, GMSI, GCIE, GCVO, KCB, DSO. IOR/L/PJ/5/396 / CC BY 4.0.

and episodes which created a new spatial order. Rather, I show how those physical and imaginative Indian spaces which Partition was designed to administer and reorder were in fact a dynamic and active component of the partitioning process. I argue that the spatial and discursive arenas in which borders are imagined, contested, drawn and enforced are deeply political sites within which both states and subjects are reconstituted.

As this introduction will show, there are already many books about Partition. Why write another? Why write it now? This book grew out of the disjuncture between, on the one hand, the often intricately detailed and theoretically complicated debates on Partition that have taken place among historians and other scholars of the subcontinent for many decades and, on the other, the relative paucity of a sustained engagement with the topic among human geographers. How could human geography, a discipline to which historical geographers have contributed so much, not have produced a community of scholars concerned with one of the most significant geographical events of the 20th century? In the past few years, historians have embraced a 'spatial turn', concerning themselves more deeply with the role that space and place play in the past and our interpretation of it, and this includes historians of Partition. Sarah Ansari and William Gould highlight the importance of place in the introduction to their 2019 book on citizenship in the newly partitioned nation-states of India and Pakistan, *Boundaries of Belonging*: 'Ignoring the spatial turn of the last couple of decades is no longer a realistic option for historians' (Ansari and Gould, 2019, p. 4). Historical geographers, similarly, are engaging more systematically with historical geographies of the 20th century, including vital work on decolonisation (Clayton, 2020; Craggs, 2014; Ferretti, 2020). Yet there remains no historical geography of the political and legal process of Partition, nor is there a geographical account of the role of geographical knowledge and practice in the making of the boundary. What would historical geography have to say, not only about Partition but about how scholars might study Partition, if its methods and modes of inquiry were brought to bear on the Partition archive and the already crowded field of Partition literature?

This book attempts to bridge this gap by recasting the material spaces and spatial practices of Partition, and by probing the ways in which space (broadly construed to encompass large-scale territorial units and small-scale rooms in courthouses, homes and offices) was both contingent and constitutive during the Partition process. Much of the literature on Partition has illustrated with great efficacy and rigour the many ways in which Partition precipitated the reconstitution of identity, very often wrought through the physical act of migration from one side of the new borders to the other, and was shot through by trauma and violence (e.g. Butalia, 1998; Khan, 2007; Menon and Bhasin, 1998; Pandey, 2001). I do not seek to challenge these powerful narratives. Rather, I argue that a more sustained engagement with the history and development of geographical practice, specifically with regard to surveying, cartography and boundary-making in the context of Partition, and of British India more generally, allows for a different way of understanding violence and trauma wrought by the Empire and by Partition. By examining the productive capacity of space, we can draw a more explicit link between the epistemic violence of Empire as articulated by postcolonial writers, and the physical and psychological trauma associated with the violence of Partition.

I show how notions of 'Indian territory' continued to develop during the later colonial period, and probe how the spatial imaginaries which underpinned British governmentality in late colonial India were refracted through Indian reformist and nationalist narratives of 'India'. I contend that these earlier spatial imaginaries informed the eventual Partition process in 1947. This historical orientation,

I argue, allows for a new mode of interrogating both the geographical and political processes at work during the informal meetings, courthouse sessions and formal negotiations involved in the boundary settlement of 1947, and opens up new possibilities for understanding the logics, techniques and exigencies that characterised the territorial re-ordering at the heart of India's and Pakistan's passage to independence. This book builds on and contributes to three distinct intellectual projects. First, the book adds a distinctly geographical approach to the Partition studies literature (a term used here as shorthand to refer to the historical, anthropological and political scholarship on the 1947 Partition of India and Pakistan), not only through its analysis of maps and geographical materials but also in its emphasis on the histories of colonial geography as technical practice and mode of knowledge production. Second, the book is the first to contribute an extended account of the 1947 Partition of India and Pakistan to the critical literature in historical geography on geography, empire and decolonisation, particularly its postcolonial theoretical framing. And third, in both its empirical focus on maps, geographical data and narratives of spatial imagining and its theoretical postcolonial framing, the book is situated within the tradition of critical histories of cartography literature, which brings postmodernist, poststructuralist and postcolonial modes of analysis to the history and practice of mapping.

The most significant empirical contribution of the book takes up the issue, identified in both scholarly and popular accounts of Partition, of the lack of geographical expertise consulted during the boundary commission hearings in Bengal and the Punjab, either in the form of an academic geographer or a government official with practical experience in boundary-making. There were, in fact, geographers who participated in the process in various ways during the summer of 1947. One of those geographers, Oskar Spate, had been employed by the Muslim League in the Punjab to assist in the development of their official arguments to the Punjab Boundary Commission. His role is acknowledged in some work on Partition (e.g. Tan and Kudaisya, 2002), and Lucy Chester (2009) makes very effective use of his personal diaries from 1947. But his geographical contribution to, and position within, the boundary commission has generally been unexplored. Intriguingly, and perhaps even more surprisingly, given Spate's illustrious career and foundational contributions to the discipline of geography, historical geographers have also overlooked Spate's connection to the Muslim League. On both counts, this is a significant oversight, and one I seek to rectify here, through the recovery of and work on a cache of previously unused archival material.

The rest of this introductory chapter proceeds as follows. First, I briefly sketch the historical background of Partition, in order to show the complexity of the story as well as the dynamic and volatile atmosphere of the final years of British rule in India. This provides historical context for the next section, where I outline the broad contours of the complex field of scholarship on Partition, highlighting the relative scarcity of interventions by geographers. This is then followed by an overview of the ways that geographers have engaged with the wider questions of empire, colonialism and decolonisation, in order to situate this project firmly in both fields. Most importantly, I situate this study of Partition within the critical history of cartography tradition, which has deep connections with the postcolonial and poststructuralist work of historical geographers and historians of mapping and empire. Throughout the book, I return to the concepts and approaches developed and refined by historical geographers of cartography and mapping in two primary ways: first, to re-read scholarly accounts of Partition through a lens of critical cartography, and second, as a lens to analyse the primary source material produced here. Finally, I set out the structure of the book and introduce the archival sources that form the basis of the empirical research presented in the book.

The Unmaking of British India

The process of independence involved the significant reordering of territory, the creation of an entirely new nation-state and the transfer of power to two new governments. The official process of Partition and the formal transfer of power was conducted within the space of two months. On 3 June, just two months before Jinnah and Mountbatten rode through the streets of Karachi, a radio announcement broadcast across India declared that British imperial rule in India would come to an end and that Independence would be granted to the Indian people (Khan, 2007). In the intervening weeks, one of the most significant geopolitical events of the 20th century was negotiated and enacted, including the making of one of the most fraught boundaries currently in existence. Scholars have, rightfully, been both fascinated and dismayed by the speed with which the borders were drawn and published, while mainstream and popular accounts of Partition often frame the boundary-making process in similar terms. 'Arbitrary and artificial – established in haste' wrote Sarfraz Manzoor in *The Guardian* in 2016. Marking the 60th anniversary of Partition, the writer Pankaj Mishra recounted in 2007 in *The New Yorker* Mountbatten's decision to 'abruptly. . . bring forward the British government's original schedule [for the transfer of power] by nine months'. Mishra sums up the takeaway from this story succinctly: 'his decision is partly to blame for the disasters that followed' (Mishra, 2007).

Despite the rushed timeline of the Partition itself, the political path to independence was much longer, and part of the wider, complex and rapidly shifting political, social and economic contexts of the time. The June announcement was preceded by a number of failed attempts by nationalist leaders and the British government to agree on a plan for independence, as well as increasing public agitation and outbreaks of violence in India. As Yasmin Khan (2007) notes, independence occurred within a post-war global context of economic hardship and political upheaval: mass food shortages and famine, the return en-masse of Indian soldiers from war-torn Europe, memories of unkept promises of independence by the British Government, and the release of Congress party leaders from prison all contributed to the rush towards a transfer of power. The Independence movement had diversified as more players entered the arena. Marxist and communist groups, Hindu nationalists, scheduled caste groups, entrenched regional political parties and others all worked to mobilise Indians and popularise ideals of independence. Meanwhile, post-war circumstances in Britain meant an overwhelming Labour victory in Westminster, led by Clement Attlee, and a sea change in Indian policy.

Politics in India were similarly shifting. There had, from the beginning of British involvement in the subcontinent, been anticolonial feeling and action among Indians towards the British. But the final stage of centralised and organised anticolonial and nationalist politics began in earnest in the 1880s, with the founding of the Indian National Congress, which was to become the political party that would assume power in India upon independence in 1947. The party of Gandhi and Nehru, the Congress from its founding had framed its mission as representing the interests of all Indians, no matter their background or identity, and insisted that, while the majority of Indians were Hindu, Congress politics were secular and the party's membership was open to all. But criticism and opposition came from a variety of positions and places. Prominent among them were the All-India Muslim League, founded in 1906 by a group of elite Muslim intellectuals and politicians predominantly from the United Provinces in north India. The League argued that the Congress' position of universal inclusion and representation could not overcome the problem of electoral democracy for minorities in India. Dalit leaders mounted a similar criticism, arguing that the Congress did not adequately represent the interests of scheduled castes. Meanwhile, Hindu nationalists, many of whom were affiliated with the Rashtriya Swayamsevak Sangh (RSS) founded in 1925, argued that Indian identity and Hinduism were inextricably linked and that India was fundamentally Hindu. The Hindu Mahasabha, a Hindu nationalist organisation within Congress, split from Congress in 1933 to become an independent political party. Meanwhile, regional parties like the Unionist Party in the large province of Punjab were suspicious of an overly strong central government, and worked to build regional alliances across identity-based communities to retain power at the regional scale.

Over the course of the first decades of the 20th century, Indian nationalist movements worked in competition and collaboration with one another, creating dynamic spaces of debate, dissent, protest and negotiation. By the mid-1940s, it was increasingly clear to the British government and the Indian parties' leadership that independence would be granted to India. The arguments honed in on what that independence would look like, and what form the new Indian government should take. Before the Second World War, the Congress Party was the most powerful of the Indian nationalist parties. In 1942, however, Congress leaders were imprisoned by the British Government over their connection to the Quit India Movement, a civil disobedience campaign for independence. During this period, the Muslim League managed to chip away at the Congress Party's hold over the Independence movement and to build a solid base of power among Indian Muslims. Jinnah, who had been working since the 1930s to rebuild the Muslim League, win seats for League candidates and create a political party with the authority to represent Indian Muslims, finally succeeded in proving to the British Government that he had a legitimate role to play in the transfer of power.

The elections of 1945 marked a significant moment in the transition process for a number of reasons. For historian Yasmin Khan (2007), the elections were the Indian people's first taste of democracy and introduction to self-rule. Shot through with ballot fraud, bribery, coercion and limited access, the elections were far from 'legitimate' by contemporary standards. But for Khan, this was self-rule in action, and it set the stage for the eventual takeover of regional assemblies, bureaucracies and institutions by the victorious parties in 1946. Additionally, the Muslim League's unprecedented and overwhelming success in the 1945–1946 elections became the most powerful card in Jinnah's hand when Britain officially withdrew from India; Jinnah had become, in Ayesha Jalal's (1994) words, 'the sole spokesman' for Indian Muslims.

Meanwhile, communal tensions continued to escalate across north India. On 16 August 1946, Jinnah and the Muslim League called for a day of 'Direct Action' where Muslim businesses would close and Muslims would gather in public places to make known their political will for the creation of Pakistan. What followed was one of the most infamous and violent series of riots in British India. Bengal in particular was caught up in the waves of communal violence (Khan, 2007). As tensions grew and episodes of violence became more systematised and organised, the British government and Indian nationalist leaders grew increasingly desperate to find a solution.

The British Government attempted, through a series of missions, to build a framework through which power would be transferred to the Indian political parties. All of these missions maintained at their core the idea of a united India. The final mission, the Cabinet Mission Plan of 1946, eventually failed when Congress, under Nehru's leadership, chose to reject its terms. Jinnah, however, was prepared to accept the Cabinet Mission Plan, sparking scholarly interest in Jinnah's true motives for supporting the Pakistan demand (Jalal, 1994).

After the 11th-hour failure of the Cabinet Mission Plan, Mountbatten was appointed Viceroy in February of 1947 and charged with the task of facilitating the withdrawal of the British Government from the subcontinent and transferring power to Indian leaders. It was decided that the boundaries would be drawn by two boundary commissions, one each for Bengal and the Punjab, which meant that when independence came, India and Pakistan would thus share not one but two international borders. The historical and geographical implications of this dual Partition have been tremendously significant. After the 3 June radio announcement, Mountbatten, Jinnah and Nehru began the process of determining the structure and remit of the boundary commissions. Each of the boundary commissions would be comprised of a chair and four Indian justices. Nehru insisted that the four judges on each commission be of 'high judicial standing', and that they should represent the primary populations and parties at play (Sadullah, 1983, p. viii). The Congress Party and Muslim League therefore were central in the nomination of each of the judges for each of the commissions. The four judges were hardly neutral players, of course, and were the ones who took the keenest interest in the way their respective parties' arguments to the boundary commissions were constructed (Chester, 2009).

Both boundary commissions, it was agreed, would be chaired by Sir Cyril Radcliffe, a highly regarded London lawyer. The commissions would hold simultaneous semi-public hearings in Lahore and Calcutta, where representatives of interested parties would be invited to submit memoranda and present arguments to the justices. One of the most curious and widely cited facts about Radcliffe was that he was not a boundary-making expert and he had not previously visited India. He was unfamiliar with the geography of this vast colonial domain and had only scant knowledge of the political and ideological battles between Indian nationalist groups that had led to the decision to divide British India into two independent nation-states. The India-Pakistan boundary still bears his name – it is known as the 'Radcliffe line' – and he has variously been understood as elusive; disinterested (both in the sense of being fair-minded and detached from colonial and nationalist emotions and intrigues); as a pawn in an Indian (specifically Hindu-Muslim) power struggle; and as an imperial bureaucrat and overseer whose influence was hedged by complex political forces (Chester, 2009).

Despite this criticism of Radcliffe's lack of geographic credentials by scholars and other commentators today, Jinnah and Nehru both consented to Radcliffe's nomination as chair, believing that his lack of knowledge and familiarity with India would make him a more neutral and objective arbiter (Sadullah, 1983). He was assisted by two secretaries, V.D. Ayer Rao and H.C. Beaumont. Both had for many years served in the Indian Civil Service and were appointed in order to provide Radcliffe with an expert knowledge of India. Although Radcliffe met briefly with members of each boundary commission before the hearings began, he chose not to attend either set of hearings as they took place concurrently, and as an effort in his mind to maintain his neutrality. He preferred to work at a distance from the cut and thrust of the legal and political debate, with arguments and evidence flown to him in Delhi after each day's proceedings. He also heard individual arguments from each judge after the hearings had finished (Spate, 1991). His award was published on 17 August 1947, but violence had already escalated in the first two weeks of August. The two-day delay in the announcement of the award did not help stem the tide of aggression and brutality. Even before the award was officially announced, the communal violence and mass migration which has become so emblematic of Partition historiography had begun. Radcliffe left the country immediately, saying he could never return and destroying all of his papers relating to his work as chair of the boundary commissions (Chester, 2009).

The intricacies and intrigues of British-imperial and Indian high politics, combined with the catastrophic violence and mass migration which attended the geopolitical act of partition, have spawned an array of interpretations and debates regarding the process and legacies of Partition from different ends of the social and political spectrum. The processes precipitating the creation of the new national borders, and the events that followed the August decision (both immediately and in the longer run), have been the subject of multiple tellings and re-tellings of Partition. These narratives and interpretations are the crux of an eclectic (Indian and Pakistani; Hindu, Muslim and Sikh; metropolitan and colonial; elite and subaltern; academic and popular) specialist literature on the subject. In recent years, this literature has played a significant role in the configuration of a wider border studies literature that envisions physical and symbolic borders and boundaries as integral to contemporary problems of cultural identity, nationhood, migration, democracy and human rights, and to configurations of sovereignty, state power and resistance in a putatively 'borderless' age of globalisation.

Historicising Partition Historiography

This section introduces the broad thematic and methodological debates in the scholarly literature on the Partition of India and Pakistan. Partition studies itself has a history. A crowded field, historical accounts of Partition have been written since the 'moment' of Partition itself. Indian nationalists and British witnesses, many of whom were bureaucrats or civil servants, wrote their own accounts of the years leading up to Partition, of the summer of August 1947, and of the aftermath. This is the period, between 1947 and roughly 1980, in which official and more traditional historical accounts of Partition were consolidated and canonised. Importantly, Partition entered official and semi-official state historiographies of India and Pakistan in relation to particular nationalist and communal ideologies and discourses and helped develop them. In many cases (and certainly in the most famous cases), these authors were current or former members of government. Eyewitness accounts from the founders and thinkers of the new nation-states, therefore, became seminal for those whose job it was to bring these newly independent countries into the international community and global economy. As early as 1949, Partition was being narrated officially by those who had observed it from various vantage points. Indian accounts included G.D. Khosla's *Stern Reckoning* (1949), V.P. Menon's *The Transfer of Power in India* (1957) and Shaista Suhrawardy Ikramullah's account of the Muslim independence movement in Bengal and the establishment of Bengal, *From Purdah to Parliament* (1963, 1998). British accounts of Partition written from the perspective of colonial officials included Penderel Moon's *Divide and Quit* (1962), H.V. Hodson's *The Great Divide* (1969), Christopher Birdwood's *India and Pakistan: A Continent Decides* (1954) and E.W.R. Lumby's *The Transfer of Power in India* (1954). The journalists

Dominique Lapierre and Larry Collins' often cited *Freedom at Midnight* (1975) recounts Mountbatten's role (largely from interviews) and positions Mountbatten's memories as a faithful and accurate narrative of Partition. And finally, one of the most significant eyewitness accounts was produced by Maulana Abul Kalam Azad, a prominent Muslim member of the Congress party during the decades leading to independence. He completed *India Wins Freedom* in 1958, but it was not made available to the public until 1988.

The proliferation of more critical academic treatments of Partition began in earnest in the 1970s and 1980s, and many of these began to move away from official and explicitly ideological renderings of their subject to foster more critical and revisionist approaches. Most prominent among these is undoubtedly Ayesha Jalal's (1994) *The Sole Spokesman: Jinnah, the Muslim League, and the Demand for Pakistan*. A path-breaking revisionist work, Jalal looked afresh at, and in some respects recovered for the first time, the complexity of Muhammad Ali Jinnah's political strategies and leadership of the All-India Muslim League. Her chiefly archival findings and argument challenged the accepted historical narrative, which held that Jinnah and the League were so intractable in negotiations that Pakistan became the only acceptable outcome. This narrative placed the blame for Partition primarily on Jinnah and his League supporters, absolving Nehru's Congress Party and Mountbatten himself of responsibility. Jalal showed that Jinnah had spent the 20 years leading up to independence attempting to consolidate a diverse and diffuse Muslim political community in India through the vehicle of a run-down and barely functioning political party (the All-India Muslim League). Even more astounding was her contention that Jinnah's demand for Pakistan was a 'bargaining chip'. She argued that Pakistan was a political card that Jinnah hoped to play in order to force the Congress' hand in agreeing to more legal and institutional protections for minorities, and, in particular, the sizeable Muslim minority, in a de-centralised but still unified independent India. The idea that Jinnah could, in fact, disagree with the demand for Pakistan and that his support for it was rooted in a more complex series of political manoeuvres, departed significantly from official narratives of the Independence movement and transfer of power. Jalal's work led the way for critical historiographical accounts of Partition that challenged colonial and nationalist narratives about the varied concepts and themes that much of the earlier work took for granted, including religious identity and communalism, hagiographies of prominent individuals and, importantly, where responsibility for Partition should be located (e.g. Ahmed, 1997; Chatterji, 1994; Page, 2002).

There has simultaneously been a proliferation of other works that depart significantly from the 'high politics' approach to Partition that imbues early and official historiography. They lean more towards regional or local politics and popular and oral histories, and emphasise the human drama and trauma of Partition. This shift away from official and institutional archival sources was subject to critique because it reframed the agenda away from identifying a seemingly more accurate and nuanced picture of how Partition came about and towards a multiplicity of irreconcilable narratives of experience rooted in memory and testimony. By the turn of the 21st century, debates about methodology as well as interpretation and narration had developed considerably (Pandey, 2001). Ian Talbot argues that much of this work grew out of the experience of 'communal violence in India during the 1980s' which 'encouraged a number of activists and scholars to draw parallels with Partition' and do so within the analytical framework and timeline of 'living memory' (2006, p. 4). At the vanguard of this movement were feminist scholars Urvashi Butalia and Ritu Menon, whose work recovered the experiences of women during and after Partition. Butalia's *The Other Side of Silence* (1998) and Menon and Kamla Bhasin's *Borders and Boundaries: Women in India's Partition* emphasised a methodology of 'history from below' (1998, p. 8). This work introduced the notion that most people whose lives were irrevocably altered by Partition had been excluded from most accounts of it. Violence and trauma now occupy a particular place in the literature on Partition. The feminist work above necessarily engages deeply with the experiences of violence, as well as the methodological challenges of articulating and narrating memories of violence. Paul Brass (2003) has written about Partition violence in Punjab in terms of genocide, as a way of understanding both the violence itself and of clarifying our understandings of collective violence.

In conversation with the feminist approaches and work focused on violence and trauma studies, there is also a strand of Partition literature that stems from the school of Subaltern Studies, which is a collective of primarily historians who formed in the early 1980s, which also attempts to historicise and spatialise the trauma and violence of this event in different and more holistic ways. In a historicist mode, subalternists like Gyanendra Pandey (2001) argue that Partition violence should not be conceived as an exceptional 'blip', or an unexpected or uncharacteristic moment in an otherwise glorious South Asian history, but as both a historical event and a collection of malleable and shifting meanings through which contemporary conflict and moments of violence in India are refracted and written into nationalist myths and historiographies. For Pandey and his colleagues, the disjuncture between studies of the high politics of Partition and the quotidian violence of Partition (and the debates over the legitimacy and quality of methodologies and sources which grow out of this disjuncture) prohibits scholars from dealing effectively with the ways in which Partition as event and Partition as memory are intertwined in processes of Indian nationalist myth-making (Pandey, 2001). He says, 'The current debate on the vexed question of memory and history, in fact, tells us more than a little about the relationship between nation and history, and history and state power' (2001, p. 7). Pandey's work is a relatively theoretical engagement with the relationship between violence, remembering and forgetting, and the implications of these for writing Partition historiography, but he points to other forms of recording memory, including oral histories, which have also been central to the empirical project of reckoning with the violence.

Reflecting David Gilmartin's (1998) call to move beyond the high politics/history from below division, much historiographical work now aims to bridge the gap between the high politics of Partition and the ways that Partition shaped the lives of people who lived through it. One of the most successful examples of such work is Yasmin Khan's (2007) *The Great Partition: The Making of India and Pakistan* (2007). Not only does Khan analyse Partition at multiple scales, she shows how events that took place in one arena had consequences in another. Ian Talbot and Gurharpal Singh's (2009) account brought into a single frame the causes and legacies of Partition, allowing for a more nuanced appreciation of Partition as a set of processes that represented both rupture and continuation. Some work has honed in on the importance of scale, moving away from the national scale to examine Partition's causes and legacies at the regional or city scale. Studies of the Punjab have been well represented in the literature (Ahmed, 2012; Chester, 2009; Gilmartin, 1988; Jalal, 1998; Singh and Talbot, 1997; Virdee, 2017). Meanwhile, Joya Chatterji (1994, 2007) challenged the dominance of the Punjab in Partition historiography by fixing her attention on the eastern border in Bengal, while Talbot and Singh (1999) have engaged a comparative approach. Sarah Ansari (2005), meanwhile, considers the legacies of Partition in Sindh, in Pakistan. Ansari and William Gould (2019) have more recently considered the significance of comparison across the border, examining how in Sindh and Uttar Pradesh, the Indian and Pakistani states were both in the process of creating new citizens. Vazira Fazila Zamindar's (2007) excellent *The Long Partition and the Making of Modern South Asia: Refugees, Boundaries, Histories* similarly illustrates how high-level attempts at state-making, through the creation of passports and the transfer of property, were both experienced and shaped by contingencies on the ground.

As in Zamindar's account, migration and mobilities have become more prominent themes in scholarship on Partition, including themes of intergenerational experiences of displacement and familial remembering and forgetting. Haimanti Roy's (2013) *Partitioned Lives: migrants, refugees, citizens in India and Pakistan, 1947–65* also examines how the state produced citizens, focusing more specifically on Bengal and the eastern border. Ian Talbot and Shinder Thandi's (2004) collection *People on the Move: Punjabi Colonial and Post-Colonial Migration* places Partition migration within the wider context of mobilities that were both facilitated and forced by colonial capitalism, Partition, and postcolonial political economy.

Many accounts of migration make effective use of oral history, demonstrating how personal testimony can illuminate the ways that everyday social lives and cultural practices were transformed by the geopolitical and bureaucratic processes taking place in government offices. Much of this work also utilises literature, art and film alongside oral history to furnish narratives of Partition that move beyond traditional scholarly modes of writing. Pippa Virdee (2013) illustrates how oral histories produce unique forms of narrating and understanding troubling and violent pasts in the context of Partition. Gera Anjali Roy and Nandi Bhatia's

(2008) *Partitioned Lives: Narratives of Home, Displacement, and Resettlement* examines displacement and the making of home in just this way. Ayesha Jalal's (2013) account of her uncle, the writer Saadat Hasan Manto's life, analyses a selection of his fiction in the context of her family archive, illustrating how Manto reflected Partition as he witnessed it in his stories. Anindya Raychaudhuri's (2019) *Narrating South Asian Partition* reads an extensive oral history collection alongside cultural representation of Partition in film and literature to examine how collective memory of Partition is formed through the imbrication of lived experience with cultural texts. He opens his book with a story about a conversation he had with his mother, declaring at the start that autobiography is central to his method, and there is a growing trend for scholars of Partition to frame their readings of Partition memory in terms of their own. Aanchal Malhotra's (2019) *Remnants of Partition: 21 Objects from a Continent Divided*, which presents memories through the telling of stories about objects that migrants brought with them, begins with an interaction with her grandmother. Neeti Nair (2011), meanwhile, reflects in her opening pages on how an encounter with a document in the archive unsettled her inherited knowledge of Partition.

There are fewer explicitly geographical and spatial analyses of Partition, although there have been some notable and highly significant interventions. Lucy Chester's (2009) groundbreaking *Borders and Conflict in South Asia* was the first archival account to take the boundary commissions themselves seriously. She argues that one of the reasons the boundary commissions have been sidelined in the historiography is because they have not been seen as critical to understanding either the causes or the outcomes of Partition. She shows how the making of the commissions, as well as the drama of the hearings themselves, were in fact contingent and did have a bearing on how the process was completed. One of her strongest contributions is her archival work on Cyril Radcliffe, showing how he might be re-inserted into the narrative of Partition in a way that neither vilifies him nor denies his impact. More recently, William Gould and Stephen Legg (2019) have edited a special collection, 'Spaces before Partition', that shows how multi-scalar historical study can help us understand how spatial reordering in the years before Partition impacted the eventual Partition itself. The collection emphasises geographical modes of analysis, demonstrating how scholars can use scale to unpack complex and often contradictory social and political boundary-making processes.

Meanwhile, a few notable and relatively isolated studies in political geography indicate that the politics of Partition are of relevance to political geographers. One of the earliest works in political geography, produced by Thomas Fraser (1984), sketched the politics of Partition in Ireland, India and Palestine, highlighting the potential for comparative studies to show how Partitions were central to British decolonisation processes. More recently, Reece Jones has argued that 'the Partition was based on the false assumption that people can be logically divided up into groups based on identity categories and territorial maps'

(Jones, 2014, p. 297). However, in framing the Partition process as 'false', he implies that there was some other 'true' reality on the ground, a reality where categories were 'blurry, overlapping and incipient' (p. 297). Meanwhile, political geographers and scholars of contemporary border studies have drawn on some of the historiographies of Partition to contextualise their varied engagements with contemporary South Asian borders (Aijaz and Akhter, 2020; Cons, 2013; Ferdoush, 2014; Jones, 2012; Jones and Ferdoush, 2018; Mustafa, 2021; Shewly, 2013; Sur, 2021; van Schendel, 2005). With the notable exception of environmental histories of water politics in the Indus basin (e.g. Gilmartin, 2015b; Haines, 2016; Michel, 1967), historical work on Partition itself has tended not to incorporate geographical modes of enquiry and analysis, while much of the political geographical work on borders remains focused on contemporary processes rather than centring historicist modes of enquiry.

This book is firmly rooted in the Partition studies literature sketched above. Informed particularly by the revisionist accounts of Partition produced by scholars such as Yasmin Khan (2007), Gyanendra Pandey (2001), Ayesha Jalal (1994), Urvashi Butalia (1998) and Lucy Chester (2009), my reading of Partition is critical in its assessment of official nationalist and imperialist accounts, while being sensitive to the theoretical and methodological tensions inherent in studying Partition as a historical, political and social process. Like these authors, I understand the violence and trauma that characterised Partition to be an inextricable part of the process, rather than as an unfortunate, or tangential, aspect of the story. And, like these authors, I am concerned with the different scales at which Partition occurred, and with recognising the importance of disentangling the drama and intrigue of high politics from the everyday trauma of individual and collective experience. Therefore, as I show in the following sections, the book critically probes the relationship between geographical expertise and scientific accuracy among the professional spaces in which territorial claims were developed and the notion that expertise and accuracy might have impacted the experiences of violence, displacement and loss felt by individuals and communities. And finally, inspired by these authors, I began my own research by attempting to uncover the territorial aspirations underpinning the idea of Pakistan. From where had 'Partition' come?

David Gilmartin (2015a) has argued in a critical review of Partition scholarship that historians and religious studies scholars have explored in detail the diversity of ways that religious thought and organisation were transformed in India in the 19th and 20th centuries, but that scholars have generally shied away from explicitly connecting these processes with the actual Partition of the subcontinent in 1947. He suggests that this is for political reasons, that, 'they have also been extremely wary of turning the history of communalism itself into a story of popular agency, for that might simply provide fodder for right-wing Hindu nationalists with their own erasures of popular agency in the name of essentialised religious – and civilisational – identities' (p. 29). Indeed, I find the possibility of reading Partition back in time to the 19th century, and of rendering it inevitable, to be both ahistorical and politically dangerous. Yet the apparent suddenness of Partition remained curious.

Such curiosity about the seeming swiftness – and to me elusiveness – of Partition heightened when I started to delve into some of the literature on Partition, especially by the tendency by historians, anthropologists and political scientists to eschew a sustained critical interrogation of the territorial and spatial aspects of the Partition process itself. Despite the fact that Partition was, at its core, a geographical process, deeply concerned with the configuration, reorganisation and reinscription of territory, the vast body of literature on Partition had generally taken concepts of territory, borders and cartography for granted, and portrayed their role as inert, focusing instead on what Gilmartin (2015a) identifies: colonial forms of knowledge production and its relationship to communalism, the reconfiguration of religious identities into political identities, and the concurrent development of anticolonial rhetorics of liberation, sovereignty and democracy. These spatial attributes and conduits of power were presented as fixed, assumed and standardised, rather than as historically and politically constructed and contested ideas and practices.

Some of the most original and enlightening books on Partition, including Khan's (2007), Jalal's (1994) and Pandey's (2001) cited above, are characterised by this geographical oversight. While they work effectively to show the contingencies and contestations of the political, social and cultural forces at work during the Partition process, they tend not to critically examine the *geographical process* in the same way. Moreover, one of the rhetorical ways in which the Partition studies literature brings its subject into focus as a historical event, and thus surmounts its fractious politics, is through the deployment of spatial and psychological metaphors to describe the psychological, affective and traumatic effects of territorial division. Common use of spatial metaphors belies a paucity of sustained critical spatial theorising and reflection in the Partition studies literature. The spatiality of Partition, and particularly issues of boundary-making, tends to be glossed rather than probed. Indeed, spatial metaphors tend to normalise and occlude, rather than open up and question, the geographical processes at work both in how the Partition was enacted and how it has been studied.

The term 'Partition' is itself profoundly geographical, and the recent spate of scholarly and creative work with titles like *Partitioned Lives*, deployed by both Nandi Bhatia and Gera Anjali Roy (2008) and Haimanti Roy (2013), and *The Partitions of Memory* (Kaul, 2001) indicates a relative consensus on the uses and meanings of spatial metaphors of 'Partition' to convey upheaval, trauma, dislocation and displacement. The metaphor of the border or boundary, exemplified by Menon and Bhasin (1998) in their book title *Borders and Boundaries: Women in India's Partition*, effectively applies the image of the new border on the map to the new social and cultural identities (for Menon and Bhasin, intersectional gendered, religious and national identities) that were forged in the wake of the making of the boundary. Such metaphors work to fix within Partition literature an understanding of Partition as a simple spatial process of territorial division and with the territories of India and Pakistan seen as sites of struggle torn asunder (in Indian nationalist narratives) or born anew (in Pakistan). The consequence of such fixed and unexamined uses of spatial concepts is that the way territorial partition works as a geographical and geo-political project and practice has not received the critical attention it warrants. While the historical and political contingencies of the partitioning process are often central to historiographical and ethnographic accounts of Partition, the geographical practices and spatial contingencies involved are left unexamined or undertheorised. Chester's (2009) work is the first sustained study that addresses the legal boundary-making process of Partition in detail. She showed how the boundary commission's structure, remit and process had an active hand in creating the eventual borders, and she effectively showed how the particularities of an individual, Radcliffe, could shape the process in both small and large ways. This book builds on Chester's focus on the Punjab Boundary Commission and extends her project by tracing the forms of geographical knowledge and cartographic materials presented in 1947 to the 19th-century colonial geographical archive. I argue, like Chester, that the boundary commission played a significant role in Partition in the Punjab, but I shift my focus away from Radcliffe and the politics of the various parties involved towards the production and mobilisation of geographical data, techniques and strategies by both the colonial government and Indian nationalists before and during the Partition process. While Chester looks in detail at the 1947 Punjab Boundary Commission, this book begins in the 19th century, examining the histories of geographical and spatial thinking in British India over a longer period.

Geographies of Empire, Colonialism, and Decolonisation

As the previous section has shown, until recently, relatively few historical accounts of Partition engaged deeply with the kinds of questions that interest critical human geographers and historians of cartography: questions, in the present context, about the relations between maps, knowledge and power; borders and boundary-making, geo-political strategies, geographical imaginations of land and territory and the mutual constitution of society and space. More specifically, the Partition studies literature has tended to skirt around the connections between the apportioning of space and the production of territorial sovereignty; and also between British colonial modes of producing geographical knowledge and controlling territory, and Indian nationalist modes of challenging that knowledge and re-shaping it for the purposes of nation-building.

Despite the salience of these geographical questions to Partition, historical geographers, past and present, had until recently made remarkably little of the territorial Partition of the subcontinent. The historical geographies of empire, and the histories of geography *and* empire, are well developed. As Dan Clayton writes in a recent overview:

> Empires have been greatly shaped by geographical understandings, traditions, practices, concepts, and institutions – of frontiers, the exotic, and the unknown; expeditionary traditions of discovery, observation, mapping, description, and classification; colonising practices of appropriation and dispossession; cosmographic and secular systems of geographical knowledge, and doctrines of environmental determinism and geo-politics; and learned societies such as the Royal Geographical Society (founded London 1830) and popular magazines such as National Geographic (founded Washington DC 1888). (Clayton, 2017, p. 1)

Critical interest in the spatiality of empire has developed within the framework of postcolonialism, which can be characterised, in broad terms, as a multifaceted disciplinary project that seeks to challenge the idea and hope that independence heralded a complete break with the colonial past and new autonomy. Rather, colonial categories, practices, power structures and ideologies of colonialism and modernity outlived the formal eclipse of empire after World War II and continued to impinge on the fortunes of the decolonised world.

Postcolonialism necessarily has recourse to the colonial past as a means of addressing undisclosed and unresolved problems in the postcolonial present, and is centrally concerned with the creation and maintenance, and resistance to, hierarchical configurations of identity and belonging – categories of 'us' and 'them', and 'inside' and 'outside'. A range of work within geography literature which might be described as a 'postcolonial geography' draws from Edward Said's well-known and hugely influential account of the power of 'imaginative geographies', and dwells on the significance of both material geographical processes of colonisation and the spatial practices of representation that shaped and supported empire (Gregory, 1995; Jazeel, 2019). Said (1979) argued that the 'Orient' was culturally produced by Western writers and artists, and that over many centuries, a Western 'imaginary' regarding the East took root which portrayed the Orient as the West's 'surrogate self' and mystical, aberrant and inferior 'other'. Said famously argued that this imaginary, this means by which the West built up an image of itself in opposition to the Orient (and by extension non-Western world), shaped and supported Western imperial expansion and colonial rule in so-called Oriental regions from North Africa to the Far East (and by extension in other areas of the world). Cultural and historical geographers have used and extended Said's insights to examine how Western European scientific knowledge (geographical knowledge to be sure, but not just geographical knowledge) produced spatial imaginaries of 'us' and 'them', 'self' and 'other', 'the West and the rest' that were often mobilised for the purposes of colonisation.

In postcolonial critique, critical analysis is often placed on text, image and discourse (or systems of representation). In this regard, particular interest has been shown in the role that potent national and Eurocentric ways of seeing distant lands and territories, and constructing the 'colonial other', played in inducing, expediting and justifying imperial expansion and colonial rule, and more specifically in how spatial practices and discourses of exploration, surveying, mapping and travel worked as tools of empire. These spaces of representation (or discursive spaces of empire) were bound up with discourses of Western philosophy, natural history and political arithmetic (economic and demographic statistics); and latterly with professional disciplines like geography and anthropology which found their *raison d'etre* in the study of exotic peoples and places, and in metropolitan spaces of the exhibition, museum and zoo, which exemplified and exuded ideas of imperial right and might, and European progress and civilisation.

Historical and cultural geographers in a myriad of ways examined how these imaginaries were built and how they operated in different ways in different colonial times, places and projects, including those of exploration (Burnett, 2000; Driver, 2001; Naylor and Ryan, 2009), communication (Ogborn, 2008), transportation (Baker, 2014); settlement and missionary endeavour (Pettitt, 2007; Wainwright, 2009); and colonial government and control (Legg, 2007, 2014). Geographers also consider how these processes and projects took place in myriad metropolitan and colonial spaces of knowledge and in a variety of documentary forms: travel narratives (Arnold, 2011; Guelke and Morin, 2001; McEwan, 2000), maps and atlases (discussed further below), photographs (MacArthur, 2022; Ryan, 2013); accounting and administrative systems; museums, exhibitions and learned societies (Bell, 1995; Driver et al., 2021; Livingstone and Withers, 2011). Historical geographers have theorised extensively about the 'othering' and fetishizing of tropical space, giving rise to a subset of literature on tropicality (e.g. Bowd and Clayton, 2018; Driver and Martins, 2005; Ferretti, 2021).

The study of geography's empire is one (albeit diffuse) strand of a much broader engagement by geographers with postcolonial theory. Tariq Jazeel's recent intervention makes two key points that are relevant here (2019). First, he reminds us that postcolonialism has itself always been geographical, pointing specifically to the ways that early postcolonial theory was preoccupied with the production of colonial and metropolitan space. In this way, he argues, geographical knowledge and postcolonialism are mutually constitutive, rather than a one-way street, where human geography engages postcolonial theory (much of which originated in literary theory and history) without actively shaping it. This partly explains why geography, as metaphor, as theatre and as a taken-for-granted problem, permeates the Partition studies literature, even as its histories and contingencies remain underdeveloped compared to writing on, for example, violence or memory. The postcolonial study of Partition is, therefore, bound up in geographical questions.

Second, Jazeel calls attention to the ways that geographers have put postcolonial theory to work well beyond the historical geographies of empire literature to show how geographers have examined issues of belonging and identification stemming from processes of decolonisation and independence. 'If colonialism was dependent upon, and productive of, the networked spatialities [of empire], then these networks and border crossings, geographers reasoned, were centrally implicated in post-colonial and diasporic lives as well as in the ways the nation-state is (re)produced culturally' (Jazeel, 2019, p. 26). The literature on geography and empire has often been concerned largely with the 18th and 19th centuries, and often with the agendas, worldviews and foibles of colonisers and the metropolitan world rather than with those of the colonised and the messy pragmatics of colonial encounter and conflict. Historical geographers have produced ground-breaking and rigorous theoretical and empirical works on the post-enlightenment promulgation of what Felix Driver (1992) termed 'Geography's Empire', but substantially more geographic research remains to be done on the post-war era of decolonisation; as Dan Clayton, who has been at the forefront of this work in geography, notes, 'remarkably little historical attention has been paid to geographers' entanglements with postwar decolonisation' (2020, p. 1543). Interestingly, geographers working within the critical traditions of development and post-development studies have studied the period of the mid-20th century, and there is excellent literature on 'cold war geographies' (Barnes, 2022; Barney, 2015; Farish, 2010), but this work has not yet been critically connected to 20th-century processes of decolonisation. Only in the past few years, more sustained attention has been paid to the geographical knowledge and practices implicated in the retreat and break-up of European colonial empires (Bowd and Clayton, 2018; Clayton and Kumar, 2019; Craggs, 2014; Craggs and Neate, 2020), and especially how colonial territories were divided and transformed into independent, territorially bounded nation-states (Clayton, 2020; Ferretti, 2020). This was a curious oversight and until recently a crucial limitation to postcolonial geography as a critical project. It also seems increasingly anomalous given the broad turn within postcolonial studies over the last 15 years or so away from the more distant colonial past and towards a concern with decolonisation as a proximate location from which to probe the phantasms and predicaments what Derek Gregory has called 'the colonial present' (Gregory, 2004).

Critical Cartography and Map Histories

Mapping and cartography occupy a central role in geography's colonial and imperial history, and historical geographers of empire often work in conversation with the interdisciplinary critical history of cartography tradition, pioneered by J.B. Harley (1988, 1989), Harley and Woodward (1987) and Denis Wood (1992), and developed latterly by Jeremy Crampton (2010), John Pickles (2004), Mark Monmonier (2006, 2018) and most especially by historian Matthew Edney (1997, 2019). The critical history of cartography project began in the 1980s, when Harley, a cartographer, began to draw on the work of Panofsky, Derrida and Foucault to reinterpret and reimagine the making and use of maps. In his 1989 essay 'Deconstructing the map', he argued that maps were not simply scientifically produced, and thus faithful (impartial and over time ever more accurate) representations of reality, but contained hidden, symbolic meanings which freighted and shaped social relations of power. Although never fully or (for many) coherently unpacked by Harley, 'deconstruction' was the approach he advocated to get at the observable and invisible (symbolic) power of maps. Influenced by Foucault's conceptualisation of knowledge as inherently power-laden, Harley had previously argued that maps are not 'inert records of morphological landscapes or passive reflections of the world of objects', but are, rather, 'a way of conceiving, articulating and structuring the human world which is biased towards, promoted by, and exerts influence upon particular sets of social relations' (Harley, 1988, p. 278). He argued that, as cartography was a form and expression of knowledge, it therefore was also a mechanism by which power could be obtained, held and exercised: 'the surveyor, whether consciously or otherwise, replicates not just the 'environment' in some abstract sense but equally the territorial imperatives of a particular political system' (Harley, 1988, p. 279). At the time, this was a profound epistemological break from the commonly held assumptions of professional cartographers. Matthew Edney notes that Harley's work was a critical response to internal assumptions and ideals within the academic discipline of cartography in the 1980s, which sought an 'epistemological ideal of cartographic perfection' (1997, p. 24).

Historical geographers have engaged with Harley's work, and with critical cartographic theory more broadly, in a variety of ways. Felix Driver has shown how central mapping was to the development of geography as a set of practices and a mode of knowledge production in the 18th and 19th centuries: the production of 'maps and charts [that] crafted a new way of seeing the worlds beyond Europe', and that this process was driven by a popular narrative of the geographer-explorer 'as a missionary of science, extending the frontiers of (European) geographical knowledge' (2001, p. 4). Scholars have also used critical cartographic analysis to challenge triumphalist Eurocentric and white settler-focused historical narratives of early exploration and discovery, recasting European scientific surveying and cataloguing exercises as part of territorial conquest (Burnett, 2000; Carter, 2010; Clayton, 2000).

Matthew Edney's (1997) *Mapping an Empire* is emblematic of this work in the context of South Asian history and has enabled much of the theoretical and empirical work set out in this book. Edney examines the role of geographical knowledge in the East India Company's campaign to discover, claim and govern Indian territory, chiefly through the implementation of the Great Trigonometric Survey and the eventual establishment of the Survey of India. Edney's intervention illustrates the spatial corollary to the more widely known statistical and enumerative strategies and technologies that were so central to Britain's colonial governmentality in India. Governmentality, articulated by Foucault most especially in his later lecture series (2007, 2008), and developed by many scholars since (see Legg and Heath, 2018), encompasses the ways in which the state exercises power across scale in order to create and maintain a productive population. This emphasis on scale makes it a particularly effective concept for doing a historical geography of Partition; governmentality simultaneously targets the individual, the family and the state, and operates through institutional frameworks, discourse and quantitative forms of analysis.

Technologies of colonial governmentality included the census, which has been the subject of extended theorisation that posits that the census, rather than capturing social and cultural identities that existed a priori, in fact was central to the production of modern identities that were politicised in various ways by colonial governance (Appadurai, 1993; Cohn, 1996; Dirks, 2001; Kaviraj, 1992; Metcalf, 1994). Ethnological and linguistic surveys created and consolidated for administrative purposes extended lists of the ethnic and racial groups that lived in British India, and the predominant languages spoken by each (Majeed, 2020). Similarly, revenue surveys were deployed for the purposes of taxation and land management (Michael, 2007). Government surveys 'improved "the exercise of power by making it lighter, more rapid, more effective," and more subtle' (Edney, 1997, p. 24). But this was always an ideal of colonial surveying rather than the reality. Edney argues that 'European states and their empires could never be so totalising. They could never be so effective' (p. 25). Maps and surveys were 'like all instruments of state power. . . exercises in negotiation, mediation and contestation between surveyors and their native contacts, so that the knowledge which they generated was a representation more of the power relations between the conquerors and the conquered than of some topographical reality' (p. 25). Imperial 'texts and maps did not present truth. . . The British simply believed that they did' (p. 26).

Even more importantly, these mapping processes were not one-way. Scholars have pointed to the ways that cartographic science and map images have simultaneously been mobilised for anticolonial agendas. Sumathi Ramaswamy's (2009) study of *Bharat Mata*, or Mother India (the divine incarnation and visual representation of the Indian nation) recovers the cartographic impulse and mode at the heart of the visual tradition of Bharat Mata – a cartographic counter-mapping from an Indian perspective – and argues that 'barefoot cartographers' both relied on and subverted the norms of cartographic practice in order to depict their spiritual and historical claim to an independent nation-state (p. 34). Meanwhile, Edney reminds us that empire was not simply created and reproduced in 'maps, subsuming all individuals and places within the map's totalising image' (1997, p. 24). He further complicates the power of imperial mapping, arguing that it 'is an ironic act, postulating as it does a double audience: the population in the mapped territories remains ignorant while another population is actively enabled and empowered to know the mapped territories' (Edney, 2009, p. 13). But, like Ramaswamy shows, this was an *ideal* of colonial cartography, rather than the reality, because colonised people were also cartographers. Indeed, colonised people made productive use of the cartographic practices and materials produced by colonisers for anticolonial ends. In *Decolonizing the Map*, James Akerman (2017) and his contributors show how decolonisation processes very often made use of colonial maps in the transition to independence, but that creative modifications could be used to decolonise national space. As Raymond B. Craib argues in his introduction, 'Maps and atlases thus helped perform the hard cultural work of decolonizing the land, the past and, in the famous phrase of Kenyan intellectual Ngugi wa Thiong'o, the mind' (Craib, 2017, p. 18, diachronic marks not in original).

Yet even as I sketch the influence of the critical histories of cartography tradition on this project, I note that, while maps form a substantial component of the archival material examined in this book, they are not the sole object of study. Nor is map analysis my chief mode of enquiry. Rather, the historical geography of Partition narrated here takes seriously Matthew Edney's recent radical suggestion that 'we should abandon the critique of maps, whether normative or sociocultural and engage instead in a critique of the ideal of cartography' (2019, p. 4). Edney argues that cartography has, among critical scholars and technical practitioners alike, been taken for granted, conceptualised as the making and study of maps. In this framework, cartography might vary across time and space but is at its core a universal practice. He points to critical work on the history of cartography, which has often attempted to bring historically marginalised forms of mapping into the fold, to recognise the diversity and multiplicity in forms and styles of cartography. Non-European, Indigenous, pre-colonial and prehistoric maps have all been studied in this way, as if they had always been examples of cartography but as a result of enlightenment and colonial modes of categorising knowledge, had been rejected as false, unscientific, inaccurate or, in the language of colonial anthropology, 'primitive' (e.g. the foundational *History of Cartography* project, originally edited by Harley and Woodward, especially Volume 2, Book 3, published in 1998). But all of this assumes that *cartography* has existed for as long as humans have bothered to represent the world in spatial and visual ways. Critical scholars have therefore focused their attention on analysing maps. Harley's foundational work is an illustrative example of this: while postmodernist and poststructuralist approaches to analysing the 'text' of the map do reveal important features that the map hides, such as power relations, ideologies, invisible or excluded perspectives and so on, cartography itself remains an ontologically stable concept, the fixed framework within which maps are critiqued.

The maps *themselves* only tell us so much about Partition, a process and event that many scholars have already told us quite a lot about. We are left with narratives that are often rigorously researched and beautifully written but remain frustratingly descriptive in some ways. This is not to diminish the fine work of scholars who have written about Partition maps, nor to claim that my own work somehow succeeds where others have not. Rather, I aim to call attention to the opportunities that Edney's new work opens up for historical geographies of Partition and to show how we might shift our critical focus away from the maps and towards what Edney calls the 'ideal of cartography' that was at work before and during the Partition process. Importantly, examining the ideal of cartography also helps explain why commentators on Partition, both scholarly and not, have appealed to geography as a form of knowledge that could have, if done properly, prevented some of the violence and trauma caused by the partitioning of territory in the subcontinent. In order to explore the ideal of cartography at work in the Partition, I bring into the frame not only the maps and the boundary-making process of 1947 but also older colonial forms of surveying and gazetteering in the 19th century. I argue that, although we cannot and should not read Partition itself into 19th-century maps of India, we can uncover the ways in which an ideal of *colonial* cartography was at work in British India, and which was imbricated in the ways that anticolonial actors, Indian nationalists, Indian and British politicians and civil servants conceptualised India and, eventually, Pakistan.

Doing Historical Geographies of Partition

As much of the above suggests, there is considerable scope to rethink the history of Partition in geographical, and specifically cartographic, terms, and open up a series of original and important questions about territory, boundary-making and the nature, meaning and productive capacities of borders. In order to tell this story, I have employed a methodology of archive-based textual and visual analysis. The book draws together and analyses an eclectic and original combination of primary documents which, when examined alongside one another, allow for the construction of a new narrative of Partition. In some sections, I use a range of archival and secondary sources to construct a narrative timeline or to reimagine the spaces and events of a historical scene. In other sections, I focus my analysis on specific significant, and usually under-examined, texts. I utilise both methods in order to more creatively and substantially highlight the ways in which spatial and geographical knowledge and practice were produced and mobilised in the late colonial period in India.

Throughout the book, I pay careful attention to the ways in which the sources presented here both produce and illuminate the discursive and material nature of borders and boundaries in the subcontinent. I show how an array of investigative projects, namely the production of atlases, geographical treatises and boundary-making handbooks, worked to produce and naturalise boundaries in colonial India rather than simply capture or reflect pre-existing physical and human areas, differences and divisions. I demonstrate how spatial boundaries and divisions in the Punjab region of colonial India were produced – rather than given – in the form of atlases, maps, diagrams, gazetteers, census reports, government quotas and reserved seats (the colonial government's policy of allocating a set number of elected seats based on identity categories), segregated schools and so forth. I also examine the question of the constitutive nature of borders and boundary-making from both 'British/colonial' and 'Indian/Pakistani' perspectives and put these categories in scare quotes to flag the notion that the investigative projects and modalities of boundary-making were implicated in creating, realising, maintaining and revising the identities and categories underpinning the colonial divide and aspirations for independence. Finally, the book aims to open up and examine the different (geographic, textual, visual) venues in which Partition took place, and how these spaces (from maps and gazetteers to census-taking and the courtroom debate) shaped the partitioning process. My re-mapping of Partition has three main geographical components: first, a regional focus on the Punjab and the Punjab Boundary Commission, in order to show how issues of scale and regional diversity both impacted and were impacted by wider national or imperial agendas before and during the partitioning process; second, the promotion of deeper historical-geographical understanding of how colonial and nationalist imaginations and constructions of the Punjab and Pakistani territory developed alongside a set of boundary-making principles and precedents; and third, a concern with how such imaginations and principles were at work in both the boundary-making practices and political deliberations that precipitated Radcliffe's August 1947 decision.

Accordingly, the book is divided into four substantive chapters. Chapters 2 and 3 deal with the 19th and early 20th centuries. These chapters work together to highlight the ways in which colonial geographical practice and imperial geographical imagination worked to construct the particular social and political conditions within which a territorial partition of India could be imagined and negotiated. They show that the possibility of Pakistan before Independence was part of a dynamic intellectual arena in which the future of India, both as a political ideal and as a bounded and discrete territorial unit, was highly contested and contingent. Chapter 2 traces the development of mapping and surveying in British India after 1858, showing first how British colonial officials produced and used geographical data to administer British India, and second how British geographers and colonial officials placed increasing emphasis on boundary-making and workable borders into the 20th century. The chapter begins with an examination of the production of the first *Imperial Gazetteer of India* (1885) by William Wilson Hunter and probes the logic of bordering at work in Hunter's mission (which has received remarkably little scholarly attention, and which was instrumental in facilitating the first major iteration of the Indian census). The chapter then traces ideologies of 20th-century boundary-making in the work of the British colonial surveyor, Thomas Holdich, and the American geographer and boundary-making expert, Stephen B. Jones, who worked for the US State Department.

Chapter 3 considers how these cartographic and surveying projects and ideologies sat alongside some of the first nationalist imaginings of 'India' and 'Pakistan'. This chapter examines the development of Muslim nationalist political and religious philosophy during the late 19th and early 20th centuries, uncovering the traces of geographical, spatial and territorial thinking that was threaded through broader nationalist discourses among Indian Muslim intellectuals. Beginning with the Aligarh school, its founder Sir Sayyid Ahmad Khan, and his idea of the two-nation theory, Chapter 3 considers how some prominent Indian Muslim thinkers and leaders appropriated, challenged and incorporated colonial geographical knowledge into their understandings of India, what it meant to be Indian, and their visions for India after the departure of the British.

These chapters set the stage for a thorough investigation of some of the key spaces of the Partition process itself in July of 1947, which form the basis of Chapters 4 and 5: the Lahore High Court, the homes and hotels where party members and officials lived, stayed and met during the course of the summer of 1947, as well as the cartographic and geographical evidence which was produced and mobilised during the Punjab Boundary Commission hearings. In Chapter 4, I undertake a textual and visual analysis of the Punjab Boundary Commission documents, particularly the hearing transcripts, which were collected over the 10-day period in the Lahore High Court, and the maps that were pivotal to the way testimony from the main parties was presented and heard. This chapter shows how geographical knowledge was one of many forms of knowledge used by delegates and members of the boundary commission to produce competing claims to territory. The chapter also draws on the material analysed in the first two chapters to illustrate how such geographical knowledge was itself power-laden and contingent, produced within and mobilised by colonial institutions in the late 19th and early 20th centuries. Finally, the chapter provides the historical context for understanding how geographers and geographical expertise contributed to the Partition process.

Chapter 5 conducts a thorough examination of unpublished and previously unused material (held in the National Library of Australia) pertaining to the boundary commission that were documents produced and collected by the British geographer Oskar Spate, who worked as an independent advisor to the Muslim League during the Punjab Boundary Commission hearings. I analyse Spate's maps, notes and reports on the court proceedings, diaries, letters and memoirs, in order to extend understanding of the geography of Partition. Particular attention is paid to the relationship between Spate's academic (in many ways scientific-rationalist) persona and his political connection to the Muslim League, and his ambiguous role as both actor in and witness to the boundary commission. In Spate's diaries, the space of the courtroom was a decidedly legal and juridical venue in which competing geographical narratives and arrangements – which were bolstered by maps, charts, tables, handbooks and manuals – were presented and debated. The space of the courtroom was an important aspect of the boundary commission hearings, lending the process an air of legality, of transparency (even in the midst of government-mandated controls on media reports), of Indian responsibility and accountability, of the possibility of compromise and negotiation.

It is equally telling that the chairman of the boundary commissions, a prominent lawyer and judge (and the only British member of the commissions), did not ever enter the space of the Indian courtroom.

In the analysis that follows, I heed the call of Sanjay Chaturvedi, who wrote 20 years ago that 'a critical geography of Partition(s) should compel us to look at the other side of the "exact" maps of the world, and to question their privileged locations, localities and directions. . . concealed behind such "settled" maps are not only the violent histories and geographies of Partition and boundary-making but also "unsettled" and "unsettling" assertions of disagreement and dissent' (Chaturvedi, 2003, p. 148). Ultimately, the book argues that Partition, as enacted by the Government of India in 1947, was not an exceptional solution to the terrifying and irresolvable problem of 'communal violence'. Nor was Partition simply the poorly planned and hastily executed result of a British government that no longer cared about India. Partition was part and parcel of imperial and colonial forms of governance, and, although the circumstances of the summer of 1947 provided a dramatic narrative, those circumstances were neither extraordinary nor exceptional. Rather, they were the product of the consolidation of colonial procedures and technologies of territorial governance carried out gradually over the course of two centuries. Crucially, the book shows how those colonial procedures and technologies are rooted in particularly modern forms of *geographical* knowledge and power.

References

Ahmed, A.S. (1997). *Jinnah, Pakistan and Islamic identity: The Search for Saladin*. London: Routledge.

Ahmed, I. (2012). *The Punjab Bloodied, Partitioned and Cleansed: Unravelling the 1947 Tragedy through Secret British Reports and First-Person Accounts*. Karachi: Oxford University Press.

Aijaz, A. and Akhter, M. (2020). From Building Dams to Fetching Water: Scales of Politicization in the Indus Basin. *Water* 12 (5): 1351. https://doi.org/10.3390/w12051351.

Akerman, J., ed. (2017). *Decolonizing the Map: Cartography from Colony to Nation*. Chicago: University of Chicago Press.

Ansari, S. (2005). *Life after Partition: Migration, Community and Strife in Sindh, 1947–1962*. Oxford: Oxford University Press.

Ansari, S. and Gould, W. (2019). *Boundaries of Belonging: Localities, Citizenship and Rights in India and Pakistan*. Cambridge: Cambridge University Press. https://doi.org/10.1017/9781108164511.

Appadurai, A. (1993). Number in the Colonial Imagination. In *Orientalism and the Post-Colonial Predicament*, edited by C.A. Breckenridge and P. van der Veer, 314–339. Philadelphia: University of Pennsylvania Press.

Arnold, D. (2011). *The Tropics and the Traveling Gaze: India, Landscape, and Science, 1800–1856*. Seattle: University of Washington Press.

Baker, J. (2014). Mobility, Tropicality and Landscape: The Darjeeling Himalayan Railway, 1881–1939. *Journal of Historical Geography* 44 (April): 133–144. https://doi.org/10.1016/j.jhg.2013.11.003.

Barnes, T. (2022). The Discipline that Came in from the Cold: American Human Geography Becomes a Cold War Social Science. *Environment and Planning F* 1 (2–4): 145–167. https://doi.org/10.1177/26349825221107646.

Barney, T. (2015). *Mapping the Cold War: Cartography and the Framing of America's International Power*. Chapel Hill: UNC Press Books.

Bell, M. (1995). Edinburgh and Empire. Geographical Science and Citizenship for a 'New' Age, ca. 1900. *Scottish Geographical Magazine* 111 (3): 139–149. https://doi.org/10.1080/00369229518736956.

Bharadwaj, P., Khwaja, A.I. and Mian, A.R. (2009). *The Partition of India: Demographic Consequences. International Migration.*

Birdwood, C. (1954). *India and Pakistan: A Continent Decides*. New York: Praeger.

Bowd, G. and Clayton, D. (2018). *Impure and Worldly Geography: Pierre Gourou and Tropicality*. Abingdon: Routledge.

Brass, P. (2003). The Partition of India and Retributive Genocide in the Punjab, 1946–47: Means, Methods, and Purposes 1. *Journal of Genocide Research* 5 (1): 71–101. https://doi.org/10.1080/14623520305657.

Burnett, D.G. (2000). *Masters of All They Surveyed: Exploration, Geography, and a British El Dorado*. Chicago: University of Chicago Press.

Butalia, U. (1998). *The Other Side of Silence: Voices from the Partition of India*. New Delhi: Penguin Books India.

Carter, P. (2010). *The Road to Botany Bay: An Exploration of Landscape and History*. Minneapolis: University of Minnesota Press.

Chatterji, J. (1994). *Bengal Divided: Hindu Communalism and Partition, 1932–1947*. Cambridge South Asian Studies. Cambridge: Cambridge University Press.

Chatterji, J. (2007). *The Spoils of Partition: Bengal and India, 1947–1967*. Cambridge: Cambridge University Press.

Chaturvedi, S. (2003). Towards a Critical Geography of Partition(s): Some Reflections on and from South Asia. *Environment and Planning D* 21: 148–153.

Chester, L. (2009). *Borders and Conflict in South Asia: The Radcliffe Boundary Commission and the Partition of Punjab*. Manchester: Manchester University Press.

Clayton, D. (2000). *Islands of Truth: The Imperial Fashioning of Vancouver Island*. Vancouver: UBC Press.

Clayton, D. (2017). Empire. In *International Encyclopedia of Geography*, edited by D. Richardson, 1–12. Wiley Online,. https://doi.org/10.1002/9781118786352.wbieg0839.

Clayton, D. (2020). The Passing of 'Geography's Empire' and Question of Geography in Decolonization, 1945–1980. *Annals of the American Association of Geographers* 110 (5): 1540–1558. https://doi.org/10.1080/24694452.2020.1715194.

Clayton, D. and Kumar, M.S. (2019). Geography and Decolonisation. *Journal of Historical Geography* 66 (October): 1–8. https://doi.org/10.1016/j.jhg.2019.10.006.

Cohn, B. (1996). *Colonialism and Its Forms of Knowledge: The British in India*. Princeton, New Jersey: Princeton University Press.

Cons, J. (2013). Narrating Boundaries: Framing and Contesting Suffering, Community, and Belonging in Enclaves along the India–Bangladesh Border. *Political Geography* 35 (July): 37–46. https://doi.org/10.1016/j.polgeo.2012.06.004.

Craib, R.B. (2017). Cartography and Decolonization. In *Decolonizing the Map: Cartography from Colony to Nation*, edited by J. Akerman, 11–71. Chicago: University of Chicago Press.

Craggs, R. (2014). Postcolonial Geographies, Decolonization, and the Performance of Geopolitics at Commonwealth Conferences. *Singapore Journal of Tropical Geography* 35 (1): 39–55. https://doi.org/10.1111/sjtg.12050.

Craggs, R. and Neate, H. (2020). What Happens If We Start from Nigeria? Diversifying Histories of Geography. *Annals of the American Association of Geographers* 110 (3): 899–916. https://doi.org/10.1080/24694452.2019.1631748.

Crampton, J. (2010). *Mapping: A Critical Introduction to Cartography and GIS*. Oxford: Wiley-Blackwell.

Dirks, N. (2001). *Castes of Mind: Colonialism and the Making of Modern India*. Princeton, New Jersey: Princeton University Press.

Driver, F. (1992). Geography's Empire: Histories of Geographical Knowledge. *Environment and Planning D: Society and Space* 10 (1): 23–40. https://doi.org/10.1068/d100023.

Driver, F. (2001). *Geography Militant: Cultures of Exploration and Empire*. Oxford: Wiley.

Driver, F. and Martins, L. (2005). *Tropical Visions in an Age of Empire*. Chicago: University of Chicago Press.

Driver, F., Nesbitt, M. and Cornish, C. (2021). *Mobile Museums: Collections in Circulation*. London: UCL Press.

Edney, M. (1997). *Mapping an Empire: The Geographical Construction of British India, 1765–1843*. Chicago: University of Chicago Press.

Edney, M. (2009). The Irony of Imperial Mapping. In *The Imperial Map: Cartography and the Mastery of Empire*, edited by J. Akerman, 11–46. Chicago: University of Chicago Press.

Edney, M. (2019). *Cartography: The Ideal and Its History*, Illustrated ed. Chicago; London: University of Chicago Press.

Farish, M. (2010). *The Contours of America's Cold War*. Minneapolis: University of Minnesota Press.

Ferdoush, M.A. (2014). Rethinking Border Crossing Narratives: A Comparison Between Bangladesh-India Enclaves. *Journal of South Asian Studies* 2 (2): 107–113.

Ferretti, F. (2020). History and Philosophy of Geography I: Decolonising the Discipline, Diversifying Archives and Historicising Radicalism. *Progress in Human Geography* 44 (6): 1161–1171. https://doi.org/10.1177/0309132519893442.

Ferretti, F. (2021). Other Radical Geographies: Tropicality and Decolonisation in 20th-Century French Geography. *Transactions of the Institute of British Geographers* 46 (3): 540–554. https://doi.org/10.1111/tran.12438.

Foucault, M. (2007). *Security, Territory, Population: Lectures at the Collège de France 1977–1978*, edited by M. Senellart, F. Ewald and A. Fontana. Basingstoke: Palgrave Macmillan.

Foucault, M. (2008). *The Birth of Biopolitics: Lectures at the Collège de France, 1978–1979*, edited by M. Senellart, F. Ewald and A. Fontana. Basingstoke: Palgrave Macmillan.

Fraser, T.G. (1984). *Partition in Ireland, India and Palestine: Theory and Practice*. London: Macmillan.

Gilmartin, D. (1988). *Empire and Islam: Punjab and the Making of Pakistan*. Comparative Studies on Muslim Societies. Vol. 7. Berkeley: University of California Press.

Gilmartin, D. (1998). Partition, Pakistan, and South Asian History: In Search of a Narrative. *The Journal of Asian Studies* 57 (4): 1068. https://doi.org/10.2307/2659304.

Gilmartin, D. (2015a). The Historiography of India's Partition: Between Civilization and Modernity. *The Journal of Asian Studies* 74 (1): 23–41.

Gilmartin, D. (2015b). *Blood and Water: The Indus River Basin in Modern History*. Berkeley: University of California Press.

Gould, W. and Legg, S. (2019). Spaces Before Partition: An Introduction. *South Asia: Journal of South Asian Studies* 42 (1): 69–79. https://doi.org/10.1080/00856401.2019.1554489.

Gregory, D. (1995). Imaginative Geographies. *Progress in Human Geography* 19 (4): 447–485. https://doi.org/10.1177/030913259501900402.

Gregory, D. (2004). *The Colonial Present: Afghanistan, Palestine, Iraq*. Oxford: Wiley-Blackwell.

Guelke, J.K. and Morin, K. (2001). Gender, Nature, Empire: Women Naturalists in Nineteenth Century British Travel Literature. *Transactions of the Institute of British Geographers* 26 (3): 306–326. https://doi.org/10.1111/1475-5661.00024.

Haines, D. (2016). *Rivers Divided: Indus Basin Waters in the Making of India and Pakistan*. Oxford: Oxford University Press.

Harley, J.B. (1988). Maps, Knowledge, and Power. In *The Iconography of Landscape*, edited by D. Cosgrove and S. Daniels, 1st ed., 277–312. Cambridge, United Kingdom: Cambridge University Press.

Harley, J.B. (1989). Deconstructing the Map. *Cartographica* 26 (2 Summer): 1–20.

Harley, J.B. and Woodward, D. (1987). *The History of Cartography*. Chicago: University of Chicago Press.

Hodson, H.V. (1969). *The Great Divide: Britain, India, Pakistan*. London: Hutchinson.

Hunter, W.W. (1885). *The Imperial Gazetteer of India*. 2nd ed., Vol. I. London: Trübner & co.

Ikramullah, S.S. (1963, 1998). *From Purdah to Parliament*. Oxford: Oxford University Press.

Jalal, A. (1994). *The Sole Spokesman: Jinnah, the Muslim League and the Demand for Pakistan*. Cambridge: Cambridge University Press.

Jalal, A. (1998). Nation, Reason and Religion: Punjab's Role in the Partition of India. *Economic and Political Weekly* 33 (32): 2183–2190.

Jalal, A. (2013). *The Pity of Partition: Manto's Life, Times and Work Across the India-Pakistan Divide*. Princeton, New Jersey: Princeton University Press.

Jazeel, T. (2019). *Postcolonialism*. London: Routledge.

Jones, R. (2012). *Border Walls: Security and the War on Terror in the United States, India, and Israel*, 1st ed. London; New York: Zed Books.

Jones, R. (2014). The False Premise of Partition. *Space and Polity* 18 (3): 285–300. https://doi.org/10.1080/13562576.2014.932154.

Jones, R. and Ferdoush, M.A. (2018). *Borders and Mobility in South Asia and Beyond.* Amsterdam: Amsterdam University Press. https://doi.org/10.5117/9789462984547.

Kaul, S., ed. (2001). *The Partitions of Memory: The Afterlife of the Division of India.* London: C Hurst & Co Publishers Ltd.

Kaviraj, S. (1992). The Imaginary Institution of India. In *Subaltern Studies VII*, edited by P. Chatterjee and G. Pandey, 1–39. Delhi: Oxford University Press.

Khan, Y. (2007). *The Great Partition: The Making of India and Pakistan.* New Haven; London: Yale University Press.

Khosla, G.D. (1949). *Stern Reckoning: A Survey of the Events Leading Up to and Following the Partition of India.* New Delhi: Bhawnani.

Kudaisya, G. (1996). Divided Landscapes, Fragmented Identities: East Bengal Refugees and Their Rehabilitation in India, 1947–79. *Singapore Journal of Tropical Geography* 17 (1): 24–39.

Lapierre, D. and Collins, L. (1975). *Freedom at Midnight*, 2nd ed. Columbia, Missouri: South Asia Books.

Legg, S. (2007). *Spaces of Colonialism: Delhi's Urban Governmentalities.* London: Wiley.

Legg, S. (2014). *Prostitution and the Ends of Empire: Scale, Governmentalities, and Interwar India.* Durham, North Carolina: Duke University Press.

Legg, S. and Heath, D., eds. (2018). *South Asian Governmentalities: Michel Foucault and the Question of Postcolonial Orderings.* Cambridge: Cambridge University Press.

Livingstone, D.N. and Withers, C.W.J. (2011). *Geographies of Nineteenth-Century Science.* Chicago: University of Chicago Press.

Lumby, E.W.R. (1954). *The Transfer of Power in India.* London: G. Allen and Unwin.

MacArthur, J. (2022). Imagining Imperial Frontiers: Photography-as-Cartography in the Mapping of Eastern Africa. *Journal of Historical Geography* 76 (April): 68–82. https://doi.org/10.1016/j.jhg.2022.03.006.

Majeed, J. (2020). *Colonialism and Knowledge in Grierson's Linguistic Survey of India*, 1st ed. London; New York: Routledge.

Malhotra, A. (2019). *Remnants of Partition: 21 Objects from a Continent Divided.* London: C. Hurst and Company.

Mansergh, N., ed. (1983). *The Transfer of Power 1942–47. Vol. 12, The Mountbatten Viceroyalty; Princes, Partition and Independence, 8 July–15 August 1947.* Constitutional Relations Between Britain and India. London: H.M.S.O.

McEwan, C. (2000). *Gender, Geography and Empire: Victorian Women Travellers in Africa.* Aldershot: Ashgate.

Menon, R. and Bhasin, K. (1998). *Borders and Boundaries.* New Brunswick, New Jersey: Rutgers University Press.

Menon, V.P. (1957). *The Transfer of Power in India.* Princeton, New Jersey: Princeton University Press.

Metcalf, T. (1994). *Ideologies of the Raj.* Cambridge: Cambridge University Press.

Michael, B. (2007). Making Territory Visible: The Revenue Surveys of Colonial South Asia. *Imago Mundi: The International Journal for the History of Cartography* 59 (1): 78–95. https://doi.org/10.1080/03085690600997852.

Michel, A.A. (1967). *The Indus Rivers: A Study of the Effects of Partition*. New Haven: Yale University Press.

Mishra, P. (2007). Exit Wounds. *The New Yorker*, August 6, 2007. https://www.newyorker.com/magazine/2007/08/13/exit-wounds.

Monmonier, M. (2006). *From Squaw Tit to Whorehouse Meadow: How Maps Name, Claim, and Inflame*. Chicago: University of Chicago Press.

Monmonier, M. (2018). *How to Lie with Maps*, 3rd ed. Chicago: University of Chicago Press.

Moon, P. (1962). *Divide and Quit*. Berkeley and Los Angeles: University of California Press.

Mustafa, D. (2021). *Contested Waters: Sub-National Scale Water and Conflict in Pakistan*. London: Bloomsbury Publishing.

Nair, N. (2011). *Changing Homelands: Hindu Politics and the Partition of India*. Cambridge, Massachusetts: Harvard University Press.

Naylor, S. and Ryan, J. (2009). *New Spaces of Exploration: Geographies of Discovery in the Twentieth Century*. London: Bloomsbury Publishing.

Ogborn, M. (2008). *Indian Ink: Script and Print in the Making of the English East India Company*. Chicago: University of Chicago Press.

Page, D., ed. (2002). *The Partition Omnibus*. Delhi: OUP India.

Pandey, G. (2001). *Remembering Partition: Violence, Nationalism and History in India*. Cambridge: Cambridge University Press.

Pettitt, C. (2007). *Dr. Livingstone, I Presume: Missionaries, Journalists, Explorers, and Empire*. Cambridge, Massachusetts: Harvard University Press.

Pickles, J. (2004). *A History of Cartographic Reason*. London: Routledge.

Ramaswamy, S. (2009). *The Goddess and the Nation: Mapping Mother India*. Durham, North Carolina: Duke University Press.

Ramaswamy, S. (2017). Art on the Line: Cartography and Creativity in a Divided World. In *Decolonizing the Map: Cartography from Colony to Nation*, edited by J. Akerman, 284–338. Chicago: University of Chicago Press.

Raychaudhuri, A. (2019). *Narrating South Asian Partition: Oral History, Literature, Cinema*. Oxford Oral History Series. Oxford: Oxford University Press.

Roy, A.G. and Bhatia, N. (2008). *Partitioned Lives: Narratives of Home, Displacement, and Resettlement*. New Delhi: Dorling Kindersley.

Roy, H. (2013). *Partitioned Lives: Migrants, Refugees, Citizens in India and Pakistan, 1947–65*. Delhi: OUP India.

Ryan, J. (2013). *Picturing Empire: Photography and the Visualization of the British Empire*. London: Reaktion Books.

Sadullah, M.M. (1983). *The Partition of the Punjab, 1947: A Compilation of Official Documents*. Lahore: National Documentation Centre.

Said, E.W. (1979). *Orientalism*. New York: Vintage Books.

van Schendel, W. (2005). *The Bengal Borderland: Beyond State and Nation in South Asia.* London: Anthem Press.

Shewly, H. (2013). Abandoned Spaces and Bare Life in the Enclaves of the India–Bangladesh Border. *Political Geography* 32: 23–31.

Singh, G. and Talbot, I., eds. (1997). Partition of Punjab [Special Issue]. *International Journal of Punjab Studies* 4 (1).

Spate, O.H.K. (1991). *On the Margins of History: From the Punjab to Fiji.* Canberra: National Centre for Development Studies, Research School of Pacific Studies, Australian National University.

Sur, M. (2021). *Jungle Passports: Fences, Mobility, and Citizenship at the Northeast India-Bangladesh Border.* Philadelphia: University of Pennsylvania Press.

Talbot, I. (2006). Introduction. In *Epicentre of Violence: Partition Voices and Memories from Amritsar,* edited by I. Talbot, 234. Delhi: Permanent Black.

Talbot, I. and Singh, G., eds. (1999). *Region and Partition: Bengal, Punjab and the Partition of the Subcontinent.* Oxford; New York: OUP Pakistan.

Talbot, I. and Singh, G. (2009). *The Partition of India.* Cambridge: Cambridge University Press.

Talbot, I. and Thandi, S. (2004). *People on the Move: Punjabi Colonial, and Post-Colonial Migration.* Oxford: Oxford University Press.

Tan, T.Y. and Kudaisya, G. (2002). *The Aftermath of Partition in South Asia.* London: Routledge.

Virdee, P. (2013). Remembering Partition: Women, Oral Histories and the Partition of 1947. *Oral History* 41 (2): 49–62.

Virdee, P. (2017). *From the Ashes of 1947: Reimagining Punjab.* Cambridge: Cambridge University Press.

Wainwright, J. (2009). 'The First Duties of Persons Living in a Civilized Community': The Maya, the Church, and the Colonial State in Southern Belize. *Journal of Historical Geography* 35 (3): 428–450. https://doi.org/10.1016/j.jhg.2008.09.001.

Waseem, M. (1999). Partition, Migration and Assimilation: A Comparative Study of Pakistani Punjab. In *Region and Partition: Bengal, Punjab and the Partition of the Subcontinent,* edited by I. Talbot and G. Singh, 203–227. Oxford: Oxford University Press.

Wood, D. (1992). *The Power of Maps.* New York: Guilford Press.

Zamindar, V. (2007). *The Long Partition and the Making of Modern South Asia: Refugees, Boundaries, Histories.* New York: Columbia University Press.

Chapter Two
Surveying and Boundary-Making in Colonial India

Introduction

In July 1947, an intellectual battle over imagined homelands and territories, waged through their geographical and cartographic representation, took place over the course of 10 days in courthouses in Lahore and Calcutta. Using geographical concepts, data and tools to justify (with varying degrees of accuracy) political and economic claims to territory, the representatives of the main parties and interested communities devised geographical arguments and presented maps to make and support their claims. Those claims, as well as the practices undertaken by the boundary commissions, were in part built on decades of geographical data collection. This chapter unearths some of the key geographical foundations upon which the boundary-making process was conducted. What older modes of bordering and boundary-making lay behind the partitioning process of 1947? What are the genealogies of the geographic data, and the cartographic images of independent India and independent Pakistan, that were presented to the boundary commissions? What techniques and methods for making borders and boundaries both developed from this colonial geography and provided a rough template for the actual negotiations and drawing of the boundaries? What is the relationship between 19th-century colonial surveying projects, colonial administrative boundary-making and Partition?

The chapter begins with an introduction to the geographies of the Punjab, with an overview of how colonial processes of administration, militarisation and

Mapping Partition: Politics, Territory and the End of Empire in India and Pakistan, First Edition. Hannah Fitzpatrick.

infrastructure projects shaped the demographics and economies in the province, both in the 19th century and into the 20th century. This focus on the colonial Punjab provides important context for the arguments that come later, in this chapter and the rest of the book. The colonial Punjab that was eventually partitioned was built during this period, and its historical geography is important to understanding why and how the partitioning process in Punjab was so fraught. Many of the territorial claims that were presented to the Punjab Boundary Commission in 1947 grappled with the geographical legacies of colonial development in Punjab. The chapter then outlines how historians of empire in India have conceptualised the British project to enumerate and categorise the Indian population for the purposes of governance and surveillance. This included primarily the Census of India, alongside later ethnological, linguistic and revenue surveys, as well as the Survey of India, which was responsible for producing maps of British India. This overview lays the groundwork for the following sections on colonial gazetteering projects in 19th-century British India, first by the English East India Company (EEIC) and latterly by the British colonial government. In these sections, I show how mapmaking, cartography and geography were central to the production and mobilisation of the colonial development of the Punjab, and that colonial surveying and boundary-making were as important as statistical enumeration to the exercise of colonial power. To examine this colonial geography, I focus specifically on the *Imperial Gazetteer of India*, begun in 1869 by William Wilson Hunter, and the ways in which this project drew on census and survey data to create visual and cartographic evidence of historical and 'natural' relationships between Indians and their territory.

The chapter then turns to the relationship between the military and administration in India, examining how colonial boundary-making was a technique for both military strategy and colonial governmentality simultaneously. Geographical knowledge in the 18th and early 19th centuries had primarily been deployed for the purposes of conquest and colonial domination, but by the time of the annexation of the Punjab in 1849, and the Government of India Act of 1858, geography had also become a tool for the consolidation of British power. The government and military in India focused their efforts on applying geographical expertise for the purpose of governance, administration and the construction and maintenance of secure and effective boundaries. In this section of the chapter, I examine one fascinating example of this relationship between the colonising forces of geography and the military in India at the turn of the century: the work of Thomas Holdich. I examine specifically how Holdich's writings on frontiers and boundaries reflect and reify many of the assumptions about territory, society and culture that appeared earlier in the *Imperial Gazetteer*.

In the final part of the chapter, I turn to practices of boundary-making in the 1930s and 1940s. By the 1940s, geographical knowledge of boundary-making was being called upon to construct new territorial orders out of older imperial models. The art and science of boundary-making, which had been central to the making and maintenance of empire, was now being mobilised to deal with rapid changes in global geopolitics. Bordering practices became one of the primary modes through which nationalist renderings of India before independence were transferred from the imagined map to the legal map. To interrogate some of the ways that geographers attempted to create a set of broadly applicable and internationally accepted norms in boundary-making practice, I examine a key text produced by the American geographer, Stephen B. Jones, in 1945. This final section unpacks how such attempts to create a standardised methodology for boundary-making incorporated and built upon older colonial geographies of borders.

I am not simply concerned with the 'conceptual' aspects of Britain's imagined India but also with the ways in which that 'image' functioned as a model and an ideal, while the everyday local practices and realities of constructing and perfecting that model betray the limits, contingencies and asymmetries between the ideals enshrined in the model and the messiness and complexities of material and practical reality.

Constructing Colonial Punjab

The Punjab came into British colonial focus in the later stages of the Raj, during the second half of the 19th century, first as a military domain and means of extending India's northwest frontier, and then also as a significant agricultural region. The Punjab was relatively unusual in this sense and made the colonial administration, territorial organisation and environmental development of the province distinctive from other regions in British India, which eventually had implications for the Partition of the Punjab in 1947. From 1849, and especially after 1858, the extension and consolidation of British colonial power in the Punjab, and India at large, involved the development of a small but influential class of professional Indians with reformist and anti-colonial inclinations and agendas that paved the way for the independence movement of the 20th century. The British intervened significantly in Punjab, and until 1947, much of that intervention had the effect of consolidating and unifying the infrastructure and economy of the province. In order to critically evaluate just how difficult and problematic the partition of the province was, it is important to understand how the Punjab came to look the way it did in 1947.

The Punjab, the 'land of the five rivers', is located in the north-western region of the subcontinent. The five rivers that give the province its name, the Jhelum, Chenab, Ravi, Sutlej, and Beas, are all tributaries of the Indus, flowing through the heart of the province, and their alluvial deposits created the plains which make up the western regions of Punjab (Ali, 2014, p. 8). To the north-east and west of the heartland lie submontane and Himalayan regions, to south Haryana, and to the east Jammu (Roy, 2011, p. 131). Punjab was the gateway between the Himalayan mountain range into the Gangetic Plain of India, and many explorers and conquerors made their way through the Punjab on their way to Delhi (Roy, 2011, p. 131). Lahore, the city at the heart of the Punjab, became a court city of the Mughal Empire when Akbar, Jahangir and Shah Jahan held court there.

During the second half of the 18th century, a series of successor states vied for control of the region, and eventually, Sikh armies began to invade, first capturing Lahore and then extending their control into Kashmir and Peshawar. In 1801, the Sikh leader, Ranjit Singh, proclaimed himself Maharaja of the kingdom of Lahore. Lahore had therefore become a significant cultural, social and economic hub for Punjabis of many religious backgrounds. Ranjit Singh expanded the boundaries of his territory and ruled until his death in 1839, at which time a struggle broke out amongst the claimants to Ranjit Singh's throne. The British took advantage of this instability and invaded the Punjab, and in 1848, officially annexed the Punjab (Talbot, 1988).

The Punjab was the last region of India to be annexed by the British and represented the last major component of the creation of British authority in the subcontinent. The province was a key component of the colonial process of bordering and bounding the British Empire in India and was vital for colonial strategic interests. As the corridor between Central Asia and the subcontinent, annexation meant sealing off and securing a major continental avenue. The Punjab was also a significant site of colonial development and consolidation of colonial governance and administration. After annexation, the Punjab made up approximately one-tenth of the area and population of British India, but its strategic and economic importance to the Raj was far greater than its relative size. Ian Talbot notes that 'From the outset, British rule was more dynamic in the Punjab than elsewhere in India' (1988, p. 10). He argues that the British had, by 1849, developed a 'mature imperial consciousness', which meant that they 'came to the Punjab to rule, not trade and were determined to transform Punjab society in accordance with the dominant early Victorian ideals of utilitarianism and evangelical Christianity' (1988, p. 10). The British implemented what came to be known as the 'Punjab School' of administration. Led by a central administration and the support of some of the most experienced officers, the British governed the Punjab through a strong and centralised military presence, which involved the instilling of a set of moral and physical standards among its officers.

The administration in the Punjab also began a series of comprehensive economic and social reforms, which included the reorganisation of the Indian army, particularly after the uprising of 1857, often called the First War of Independence, and which the British government opportunistically and inaccurately blamed on the Indian army at the time. The Punjab emerged as 'the main military labour market of the army in India during the second half of the 19th century', due in part to the loyalty of much of the Punjab's population during the uprising, and in part to the development of the British colonial ideology of race and difference which constructed the Punjab as the home of the 'martial races' (Tan, 2005, p. 33). This led to the creation of what Tan Tai Yong has called the 'garrison state' (2005).

Entwined with this consolidation of military control was the reshaping of agricultural and land ownership policies in Punjab, which were central to the British revenue extraction and state-building agenda in the region by the second half of the 19th century. The further development of the irrigation system in the Indus Basin region, first through the maintenance and expansion of the existing canal system, and then, by the 1880s, through the building of new canals would 'transform the Indus landscape', as David Gilmartin puts it (1994, p. 1132). Gilmartin argues that, while the British government of India saw its new canal system as a magnificent feat of engineering, the canals were also central to the process of state-building in the province. The engineering and construction of the canal system itself was part of a wider transformation in social and economic life in the Punjab. Large tracts of land in the western regions of the province that had once been too arid for large-scale agriculture were now available for farming. The government strictly controlled land use and ownership in these new canal colonies, through awarding grants to specific grantees or allocating land for specific uses. The canal colonies grew rapidly as people in Punjab began to move from the eastern regions of the province to the west, where employment opportunities were more readily available. The canal projects in British-administered Punjab, as Imran Ali notes in his foundational history of Punjab during this period, facilitated 'the greatest expansion in agricultural production in any part of South Asia under the British' (2014, p. 10). The social and economic effects of this rapid large-scale migration within the province eventually made the geographical realities of territorial partition in 1947 particularly difficult to map.

Counting, Surveying and Mapping India

The large-scale development of the Punjab both relied on and reinforced the need for the colonial regime of knowledge production in India. As the EEIC, and latterly the British government, gained territorial and political control over greater swathes of India during the course of the 18th and 19th centuries, the need for a more comprehensive and systematic knowledge of the subcontinent grew (Bayly, 1993). In this section, I begin by introducing the census and the Great Trigonometric Survey of India, before outlining the key ways that statistical surveying and geographical surveying were entwined ideological projects. This is then followed by an overview of the key gazetteering projects that led up to the establishment of Hunter's *Imperial Gazetteer of India* project.

The historiography of the census is well-trodden ground, but I review some of it briefly here in order to bring sections of the colonial geographical archive from this period into its framework. Historians, including Ranajit Guha (2003),

Gyanendra Pandey (2012), Nicholas Dirks (2001), Sudipta Kaviraj (1992), Arjun Appadurai (1993) and Bernard Cohn (1996) have written extensively about the ways in which imperial technologies constructed British India, rather than simply recovered and described some pre-existing Indian reality, and how such technologies were pivotal to the way India was and could be governed. Considerable critical attention has been bestowed on the census, perhaps because it provides the most comprehensive and illuminating evidence of the colonial assumptions and logic at work in the categorisation of India and Indians. But how do we go about drawing out the specifically spatial nature of the census and other colonial practices of data collection? One way is to bring into a single frame the historical geographies of the surveying and gazetteer projects alongside the Indian censuses of the 19th century, which I sketch out here in order to show how spatial and cartographic representations of population data from the later 19th century were caught up in the wider political and intellectual debates that framed Partition in 1947.

The census was profoundly geographical, operating at a variety of scales and through different political modalities, and Indian intellectuals as well as British administrators grasped that a significant part of the power of the census lay in its ability to reach across, connect and perform different kinds of work at local, regional and sub-continental scales. The census was not simply an exercise in colonial power over passive colonised peoples, of course. Kenneth Jones (1989), in his work on Hindu modernising movements in 19th-century Punjab, argues that much of the power of the census (and, by extension, that of the gazetteers) lay in the fact that Indians, rather than just the government, used colonial administrative data to articulate Indian identities, Indian pasts, and Indian culture in ways that the government did not necessarily anticipate or desire. He shows how Hindu religious and political leaders began to use the census as a means for understanding both the strength of the Indian Hindu community as well as the relative success of Christian missionary projects in gaining converts among Hindus. The census became the litmus test whereby religious communities were able to assess their social and political power relative to other communities in India. Intertwined with this counting game was the development of a political system which incorporated and exploited the Indian professional classes. Such a system, designed on Western modes of governance, relied on representative politics, which takes as its basis the distribution of the population across territory. The political system that evolved in India therefore grew out of and reinforced colonial ideologies of identity and political subjecthood which were encapsulated and shaped by the government's program of population surveys.

During the same period, EEIC administrators were also attempting to standardise and consolidate the various geographical surveying projects taking place in its Indian territory. This was because geographical knowledge – of climate, relief and vegetation, and the density and distribution of population, resources, economic activity, customary forms of land-holding and government and ethnic and political rivalries – was key to the government's ability to collect taxes and administer and police its expanding and diverse territory. Colonial gazetteers, examined in the next section, became the most comprehensive mechanism by which the government of India could insert its understanding of the Indian population into its already comprehensive and long-standing geographical archive.

That geographical archive came to be centred around the Survey of India, which provided much of the geographical information for the gazetteers of India explored in detail in this chapter. The Survey, which began in 1767, grew out of a perceived need on the part of the British government in India to draw a single map of India in its entirety. Prior to the Survey, the EEIC had been collecting geographical data on military expeditions and as scattered smaller-scale projects, but these surveys were inconsistent, lacking in standardisation, and were extremely expensive. It was not until James Rennell (the Surveyor General of Bengal and, later, a founder of the Royal Geographical Society) compiled this data in 1765, in the first regional survey of Bengal, that a more comprehensive British map of Indian territory was constructed (Edney, 1997, p. 17). The Survey of India was the next step in this process of standardising and centralising the surveying and mapping of the subcontinent.

In the 19th century, the Survey of India, along with various revenue and cadastral surveys, was developed beyond its original purpose of mapping the entirety of the subcontinent and was deployed primarily to deal with a perceived lack of knowledge of Indian systems of landownership, boundaries and population. The Great Trigonometrical Survey (GTS), a project that began at the end of the 18th century and extended throughout the 19th century was central to this project. Matthew Edney (1997) illustrates how the GTS generated a vast amount of cartographic data on India and served as the basis for the cartographic imagination and realisation of what the British believed to be a unified, and effectively controlled India, and 'presented. . . a single and coherent view of South Asia' (Edney, 1997, p. 25). This was part and parcel of what Edney has identified as the British colonial attempts to construct a 'unified conceptual image' of India.

Edney argues that the purpose of reorganising the survey offices and consolidating all survey projects under a Surveyor General's office was to create a 'geographical archive' characterised by a single authoritative and consistent map of India. Such a map created the visual and geographical foundations for the implicit claim on the part of the British that India was, in fact, a unified territorial unit. Perhaps most relevant here is Edney's claim that India was only such a unified territorial unit in the context of *British India*; while the British surveyors and government officials may have believed that they were 'discovering' and 'recording' the geography of 'India', they were, in fact, constructing and shaping a new colonised 'India' which should be understood specifically as *British India*.

Bernardo Michael notes, 'company officials possessed scant information about the social, political, economic and cultural aspects of the lives of the people in these territories. . . and colonial officials were hard pressed to discover the boundaries, internal divisions and organisation of their dominions' (2007, p. 78). Scientific modes of surveying and cartography were one of the most important mechanisms for asserting control over territory within the British Empire. Surveys enabled the demarcation of boundaries on the ground, while maps rendered and fixed those boundaries visually. The spatial work of colonial surveys was as important as the categorisation work in shaping the Raj. The empire was territorially constructed through the seemingly objective scientific process by which two-dimensional cartographic representations of territorial units and the boundaries that marked them were made visible. Not only did such boundary-making aid in the process of 'constructing' British India, but this sort of panoptic visualisation allowed for more precise surveillance and governance – at least in theory. The metaphor of the 'panopticon prison', which Foucault famously drew from the liberal English philosopher Jeremy Bentham, works, albeit unevenly, here to describe the surveillant and visual capabilities that the British ascribed to the map (Edney, 1997; Foucault, 1977).

The maps produced by the British government in India throughout the 19th century were designed primarily as tools for governance, to be used by colonial officials on the ground for the purposes of tax collection, police administration and law enforcement. In particular, revenue and cadastral surveys were used to develop a baseline for understanding local Indian modes of organising territory and property, and, eventually, to reorganise the older Indian boundaries into new divisions that were more easily represented cartographically. Much of the territory to be surveyed was difficult to map; often, districts and villages (called *parganas*) which had previously been administered by the Mughals did not consist of contiguous units of land, but, rather, had many detached and separate segments which crisscrossed, overlapped and generally left British surveyors scratching their heads (Michael, 2007).

The revenue surveyors tended to solve this perceived problem of contiguity simply by redrawing the map, and then demarcating the new boundaries on the ground, setting up police districts (*thanas*) to coincide with the new boundaries and moving all records pertaining to transferred *parganas* to new central offices. In this way, the British government laid the foundations for the new cartographically based geographical organisation of India, rooted in contiguous territorial units bounded by clearly demarcated boundaries which corresponded to the social, political and cultural organisation of the people who lived there (Michael, 2007). All of these projects were precursors to the eventual project, undertaken by William Wilson Hunter, to compile, condense and map these disparate datasets in order to make them accessible primarily to colonial administrators but also to the British public.

The imperial and provincial gazetteers of India were, in important respects, the apogee of these enumerative (census) and cartographic (survey) endeavours, bringing population statistics and survey data together to represent the spatiality and geographical differentiation of the Indian population. The spatial imaginary of colonial control over India required the entwining of well-studied colonial identity categories with geographical and survey data. Many of these categories were filtered through colonially constructed knowledge binaries for the purposes of culturally, socially and administratively categorising and 'othering' Indians: the binaries of science/tradition and secularism/dynamism were crucial for the colonial construction of modern identity categories like religiosity, caste, ethnicity and language group. The spatialising process, whereby such identities were imaginatively and cartographically linked, either through theories of environmental determinism or through administrative surveillance practices like the census, created a particular spatial imaginary of India through which the British could govern India.

Histories of Gazetteers in 19th-Century India

While the Indian census and the Survey of India have been examined in critical detail, colonial gazetteers have received remarkably little attention. This is especially interesting, given that the first gazetteer was influential in the conducting of the 1881 census (McDermott et al., 2014). In one of the few comprehensive surveys of atlases and gazetteering in South Asia, Henry Scholberg highlights their wide-ranging aims: 'A district gazetteer is a comprehensive description of a district or state of British India, published privately, in series, or under the auspices of a governmental body, and including historical, archaeological, political, economic, sociological, commercial and statistical data' (1970, p. iv). Scholberg notes that they were primarily designed to 'provide information on their districts for incoming district commissioners and other local officials, as well as to inform the general public' (p. v).

The gazetteers and their related documents are perhaps better understood as manuals or guides, rather than as encyclopaedic collections of data pertaining to British India, although Hunter's gazetteers, upon which I base the bulk of my analysis in this section, allow for a more nuanced understanding of the role these texts played in the construction and reproduction of colonial forms of knowledge. Importantly, and despite the warnings of postcolonial historians that the colonial archive should be read, as Walter Benjamin (1969, p. 257) said, 'against the grain of history', much of the secondary literature on the British Empire in India, including much of the otherwise critical work I rely on here for historical context and narrative, continues to reproduce and redraw the images and data which were first published in the gazetteer itself. There is relatively little critical discussion or engagement with the contexts (of militarisation, of demographic detection) and the practices within which the gazetteers specifically were produced. David Arnold, however, has written briefly about the role of the gazetteers in the wider context of the proliferation of 'state-run' and 'politically dependent science' after 1858 (2000, pp. 130–131). Arnold observes that the gazetteers 'aspire to be more than a compendium of place names and geographical descriptions. Its many volumes constituted a systematic and consciously scientific ordering of information about districts, provinces and an entire empire' (pp. 130–131). Importantly, they were produced not by scientists but by bureaucrats and ICS officers, and represent, therefore, 'the empire's view of science rather than the scientist's view of empire' (p. 132).

Early Gazetteers

Hunter was not the first to conceptualise a gazetteer of India. In fact, one of the first references by the East India Company to the idea of a gazetteer appeared in 1803. A letter to relevant departments called for the collection of 'such information on the Chronology, Geography, Government, Laws, Political Revolutions and progressive stages of the fine arts, and particularly on the former and present state of Internal, and Foreign Trade' for the purposes of compiling a 'General History of the British affairs in the East Indies' (Scholberg, 1970, p. 1). Four years later, Dr Francis Buchanan was assigned the task of undertaking a statistical survey of British territories in Bengal, then called the Presidency of Fort William. This was one of the first attempts at linking statistical data about the Indian population with geographical data in order to develop a *historical* account of Britain's activity in the region. Walter Hamilton's 1820 *A Geographical, Statistical, and Historical Description of Hindostan and the Adjacent Countries* was one of the first attempts at an all-India gazetteer.

Many district gazetteers were attempted, with varying degrees of success, in the regions of India where the British were most embedded: the Bengal and Madras presidencies. As Scholberg notes, it is unsurprising that there are more statistical and geographical accounts of Bengal than any other province, for the region was the first in India where the East India Company generated a consolidated system of control (1970). As the Punjab, including the regions that became the North-West Frontier Province and Delhi, were fully annexed in 1849 and incorporated into British Indian territory, the East India Company's forms of governance and control moved away from acts of conquest to processes of administrative consolidation. One of the most important aspects of this latter process of consolidated colonial governance was the collection, systematisation and distribution of all available colonial knowledge on India and its population.

The East India Company had long been in the business of knowledge production, of course, and as Hunter acknowledged in his preface to the first edition of the *Imperial Gazetteer* in 1881: 'many able and earnest men had laboured at the work, manuscript materials of great value had been amassed, and several important volumes had been published. But such attempts were isolated, directed by no central organisation and unsustained by any continuous plan of execution' (Hunter, 1885 p. vii). If, as Edney argues, the late 18th-century and early 19th-century GTS and associated processes of geographical data collection had *constructed* the territorial-political unit we call 'British India', then during the second half of the 19th century renewed efforts at geographical knowledge production were put to work *consolidating* the administrative bureaucracies required for the maintenance of a more expansive and knowledge-hungry colonial governmentality in India.

Such a shift in knowledge production was part and parcel of a more general move towards generating a totalising and all-encompassing knowledge of India which could be written and indexed *within a single authoritative source*. In practical terms, this meant bringing previously unconnected and separate forms of administrative, historical and geographical data into a single – totalising and teleological – narrative of India's people and territory, and past, present and future, and with the arrival and consolidation of British colonial state in India represented as the apogee of Indian progress. It was believed that such a narrative would be essential to the smooth governance and management of India; it was certainly crucial to making British rule appear legitimate and progressive. Dirks (2001) writes that the gazetteers played a significant role in the establishment and practice of governmentality, particularly in their reliance upon and mobilisation of the census.

To this end, in 1854, Edward Thornton, commissioned by the directors of the East India Company, compiled *A Gazetteer of the Territories under the Government of the East-India Company and the Native States on the Continent of India*; a second edition of the book was published three years later for the purpose of disseminating 'the great mass of information comprehended' on India to 'the British public in a cheap and convenient form' (preface, no pagination). Thornton's gazetteer was one of the first attempts to compile a gazetteer covering the whole of India. Thornton's goal was twofold:

> 1st, To fix the relative position of the various cities, towns and villages, with as much precision as possible, and to exhibit with the greatest practicable brevity all that is known respecting them; and 2ndly [sic], To note the various countries, provinces or territorial divisions, and to describe the physical characteristics of each, together with their statistical, social and political circumstances. (Thornton, 1857, no pagination)

Thornton's gazetteer heralded a shift in Company thinking: while previously knowledge about India was often collected for specific and distinct purposes across different scales, by the 1850s, the Company had explored and annexed a large area of Indian territory, thus requiring a more comprehensive and standardised system of knowledge production and dissemination. Thornton wrote in the preface to the 1857 edition that the book included, 'minute descriptions of the principal rivers and chains of mountains; thus presenting to the reader, within a brief compass, a mass of information which could not otherwise be obtained, except from a multiplicity of volumes and manuscript records' (no pagination). The gazetteer was, he suggested, the most comprehensive source of its kind to date. Thornton claimed that his gazetteer 'may be regarded as an epitome of all that has yet been written and published respecting the territories under the government, or political superintendence, of the British power in India' and that it was considered to be 'a complete history of India, untainted in any degree by political bias' (no pagination).

Thornton's gazetteer pushed the Company's ethos and ideology away from brash commercialism and towards more liberal tenets of imperial beneficence, and it allowed the Company to 'stand back' and look at itself and the diverse colonial districts and commercial activities over which its agents ranged. At the same time, there was a certain idealism in Thornton's field of vision. For the work that went into his gazetteer was compromised, unsurprisingly, by the same problem that plagued earlier attempts at manuals and handbooks on India: the lack of accessible and consistently accurate data for the whole of India, and the lack of a consistent and uniform system by which that data was collected and interpreted.

Soon after the second edition of Thornton's gazetteer was published, the British Parliament passed the Government of India Act, which dissolved the British East India Company and transferred its duties and responsibilities to the British Crown. The Government of India expanded the East India Company's project of collating and systematising knowledge about India, culminating in the efforts of William Wilson Hunter. Hunter arrived in India in 1862 after joining the Indian Civil Service the year before. He was posted to Birbhum district in Bengal, where he became interested in the history and local traditions of the region. His first book, a dictionary of Indian languages, was published in 1868, with two more completed by 1871, and he is perhaps most well-known among scholars of South Asian history for his 1871 book *The Indian Musalmans: Are They Bound in Consciousness to Rebel Against the Queen?*, a racist and Islamophobic account of the role of the North Indian Muslim community in the 1857 uprising. Hunter had been tasked with devising a statistical survey of India, which would become the foundation of his gazetteering project. As Arnold noted, Hunter was a civil service officer with interests in statistics and history, but he was more of a compiler than a researcher, and as such, his work was designed primarily to synthesise knowledge of India with knowledge of administration and management for the purposes of improving British governance in India.

Geography in the Service of Empire

In this section, I turn to Hunter's *Imperial Gazetteer of India*, and I show how these gazetteers represented colonial ideals of accuracy, standardisation and accessibility, ideals which required systematic data collection practices. I then show how these ideals were never fully achieved, arguing that some of the authority inscribed in the gazetteers was produced not through accurate data but through the reproduction of claims by external sources, regardless of the quality of the original information. In other words, geographical authority was eventually ascribed to Hunter's work *because* it was Hunter's work.

In the preface to the first edition, Hunter wrote, 'The ten years which followed the transfer of the government of India to the Crown in 1858 produced a new set of efforts towards the elucidation of the country. . . the controlling power in England had now passed from a body of experts, the Court of Directors, to Parliament and the nation at large' (Hunter, 1885 pp. vii–viii). This statement reflected a key aspect of his overall vision for his project: to take the collection, management and dissemination of data out of the hands of individuals and distribute it systematically and publicly. He also had some new, relatively more standardised data to work with, collected through the first modestly successful all-India census, taken in 1871, and on survey maps provided by the Survey of India office. Importantly, Hunter's gazetteers and the Indian census, in its post-1857 form, were entwined projects from their inception, sharing staff as well as data. H.H. Risley, best known for his work on the ethnographical surveys of India and his role in the codification of caste and race, and on the Indian census, worked closely with Hunter as an administrator on the first edition of the *Imperial Gazetteer of India*, and eventually became one of the editors of the project after Hunter's death in 1900. Under Hunter's leadership, the project did manage to grow, both in scope and in published output. The first edition comprised 14 volumes. The first four volumes provided an overview of Indian history and geography, while the accompanying 10 volumes presented specific data and were organised alphabetically by place name. Subsequent editions, published after 1901, were expanded to 26 volumes.

Hunter envisioned a large-scale project that would include the creation of two of the most effective and powerful means of assembling and disseminating population data and geographical information about India: through a statistical survey and a gazetteer. Hunter and the Governor-General, Lord Mayo, believed that 'accurate and accessible information regarding India was to become, under the new system, an essential condition for the safe exercise of that control' (Hunter, 1885, p. viii). The gazetteers would be a compilation and consolidation of the geographical, geological, ethnographical, linguistic, meteorological, and population data. Hunter, in agreement with the Government of India, insisted that a useful and effective gazetteer should be built on uniformity, consistency, and efficiency. Hunter's plan 'endeavoured, first, to eliminate the causes of previous failures, by providing a uniform scheme, a local mechanism, and a central control, second, to clearly define the objects of the present undertaking' (Hunter, 1885, p. x). Hunter was tasked by the Viceroy to 'submit a comprehensive scheme for utilising the information already collected; for prescribing the principles' to be thenceforth adopted; 'and for the consolidation into one work of the whole of the materials that may be available' (Hunter, 1885, p. ix).

Hunter, like Thornton before him, emphasised his aspirations for a complete and total narrative of India: 'For here I endeavour to present an account, which shall be at once original and complete, of a continent inhabited by many more races and nations than Europe, in every stage of human development, from the polyandric tribes and hunting hamlets of the hill jungles, to the most complex commercial communities in the world' (Hunter, 1886, p. vi). His endeavours were clearly linked with what Cohn has described as the 'encyclopedic quest for total knowledge' that characterised the Victorian age (Cohn, 1996, p. 8). Unlike Thornton, Hunter managed to design and execute a more successful and authoritative model (at least according to the standards of the Raj) for a gazetteer of India, a model which formed the basis for the Government of India's geographical knowledge production throughout the 20th century. Of course, Hunter's project was emblematic of the colonial government's drive to construct a totalising narrative of Indian history which could effectively incorporate both historical and moral justifications for the continued imperial domination of India by the British. The success of the gazetteers must be understood in this context, and they were, therefore, an effective geographical contribution to the official British narrative of India contained within the colonial archive.

Hunter was appointed director-general of statistics for the Government of India in 1871, a position which allowed him to compile district-level statistical accounts while compiling his *Imperial Gazetteers*. He observed that previous attempts at compiling gazetteers had failed for two primary reasons: 'from want of local organisation' and 'from want of central control' (1885, p. ix). In order to deal with both issues, he encouraged uniformity in the data collection processes: 'With a view to securing uniformity in the materials, I drew up six series of leading questions, illustrating the topographical, ethnical, agricultural, industrial, administrative and medical aspects of an Indian District' (p. xi). He also emphasised efficiency and punctuality, noting, 'Provincial Compilers were appointed, each responsible for getting in the returns from the District Officers within the territories assigned to him; for supplementing those returns by information from heads of Departments and other local sources; and for working up the results' (p. xi). This process, Hunter argued, would result in the collection of 'the best local knowledge. . . on a uniform plan and within a reasonable time' (p. xi). Perhaps most importantly, the entire project was supervised by Hunter himself, who was responsible for the supervision of both the processes of data collection and the dissemination of results. Hunter claimed that his gazetteer provided the necessary centralisation of all stages of the project, which was essential if he was to deliver a cheap, useful and timely publication.

Hunter's analysis of racial groups in India neatly illustrates his approach. Hunter was concerned with the spatial distributions of the Indian population and used historical narratives of migration to locate and map perceived racial and tribal differences upon the British survey of India. Indians had their own conceptions of difference, of course, and there were spatially organised identities among the Indian population before the British began their surveys in the 19th century. The Mughal Empire, for example, had developed its own method of mapping revenue districts for the purposes of tax collection, which the British inherited when they annexed Mughal territory. However, the British method was unique, as was Hunter's later, in the way it utilised counting and numbering in order to categorise the Indian population using a system of classification derived from British renderings of Indian territorial history and British ideas about 'race'. It was this system of counting and numbering that was central to colonial systems of order and governance, and to Hunter's gazetteer project (Appadurai, 1993). According to Hunter, the Statistical Survey and the *Imperial Gazetteer* served two primary purposes: 'for the use of Indian administrators and for the use of the Controlling Body in England' (p. xv). The combination of signifiers in the project – imperial and gazetteer – captured the way Hunter hoped his work would be useful to both metropolitan-imperial and Indian-colonial officials. He noted, however, that a third purpose was 'for the use of the public', and it was primarily for this purpose that the Statistical Survey was condensed 'to a practicable size for general reference' (p. xv).

The gazetteer quickly became the authoritative text upon which officials in India (and others) based their knowledge of the country, and the project became a sort of umbrella for smaller scale regional gazetteers. Benjamin Lewis Rice, director of the Mysore Archaeological Researches, for instance, wrote in the preface to the first edition, published in 1876, of *Mysore: a Gazetteer Compiled for Government*:

> When, in 1874, Dr. Hunter, Director-General of Statistics, who is charged with the editorship of the Imperial Gazetteer for the whole of India, visited Bangalore, I was able to lay before him the plans I had formed for the work, and at his request undertook to prepare for Mysore a manual of each District separately, which I had not at first intended. . . I am now glad that I did so, as it obliged me to go more minutely into several subjects. Dr. Hunter again paid a visit to Bangalore in January 1876, when a part of the work had been printed, and in his report to Government was pleased to express the strongest approval of what had been done. (reproduced in Rice, 1897 p. xiii)

Indeed, some of Rice's source material was eventually revised and published in subsequent editions using Hunter's data, effectively cementing the authority of the gazetteer, and, importantly, the census and survey data which provided the bulk of statistical and geographical information included in the *Imperial Gazetteer*. Importantly, the third edition of Thornton's *Gazetteer of the Territories under the Government of the Viceroy of India* (which provided a foundation for the initial gazetteers, and was itself a source text for Hunter, particularly for material on the princely states) was published in 1886, and a significant portion of updated material it included was drawn from Hunter's *Imperial Gazetteer*:

> As regards the numerous alterations and additions that have been made, the Imperial Gazetteer has of course been the main source of information, and has been freely used; the Editors desire to acknowledge the great advantage which they, in common with all that section of the public that is interested in India, have derived from the completions of this great national undertaking. (Thornton et al., 1886, p. viii)

We see here the ways that a relatively small selection of reference texts drew upon each other in order to construct a circular form of authority, and how the reproduction of older data alongside newer data in each successive edition cemented the perceived authority of the information contained within them.

Hunter also emphasised that much of the authority and usefulness of his gazetteer stemmed from his consistent judgement about, and assessment of, which places and sites should be included. Previously, he observed,

> In the absence of systematic materials, [gazetteers] had to depend on the chance topography of tourists, or on a place happening to find its way into the records of the India House. A petty hamlet in which some traveller had halted for a night, or any locality which had formed the subject of a correspondence with the Court of Directors, stood out in bold relief; while great tracts and rivers, or the most important features of large Provinces, were passed over without a word. My first business, therefore was to take care that every place which deserved mention should be enumerated; my second, to see that it received neither less nor more space that its relative importance demanded. (Hunter, 1885 p. xvi)

Hunter saw his work as evidence of scientific progress in India – as illustrating how older modes of data collection and scientific measurement employed by the government lacked accuracy and rigour. His statistical survey and gazetteers were, he insisted, a necessary and valuable improvement upon prior, scattered and inferior modes of research. Such messages, about the substitution of science, systematicity and professionalism for amateur knowledge and local know-how, were an important characteristic of late 19th-century empire-building generally. Hunter's emphasis on objectivity and uniformity in the construction of what was an inherently subjective (indeed, biased) process of ordering and categorising underpinned the connection between science and empire. He emphasised his commitment to comprehensive and complete compilation; his gazetteers attempted to present the most recent information on India in an easily accessible format. Such a complete and comprehensive account was ideal for the colonial state, designed as a means of exercising official knowledge for the sake of gaining and wielding power. In reality, of course, such an ideal narrative never managed to reflect reality, and the ideal of an ahistorical and fixed totalising narrative was entangled with the messy reality of a dynamic and shifting colonial Indian society.

Cohn has identified a similar shift – emphasis on totality and systematicity – in the production of historical knowledge during this period (Cohn, 1996, p. 5). Yet such totality and systematicity remained elusive because Hunter was never able to overcome a fundamental obstacle to his goal of a complete and total account of the statistics and geography of India in the *Imperial Gazetteer*: much of the subcontinent was not directly ruled by the British. The princely states, which Hunter called the 'Feudatory' or 'Native' states, proved more difficult for him to measure for a variety of reasons, which he outlines in his introduction to the *Imperial Gazetteer*. Ultimately, 'the confidential relationship between the Government of India and its Feudatory States, the dislike of the native Princes to inquiries of a social or economic character, and the scrupulous delicacy of the Foreign Office to avoid grounds of offence, have rendered a complete treatment of such territories impossible' (1885, p. xiv). As a result, Hunter was forced to reproduce data on the princely states already included in Thornton's 1854 gazetteer (published over 30 years prior).

The gazetteers produced during the 1880s marked a significant shift in colonial modes of knowledge production. The volume of data collected and published under the auspices of the gazetteer project was extensive. Due to the contradictory nature of the project, which emphasised both a complete and comprehensive collection of all available information on India on the one hand, as well as a concise and useful presentation of that data on the other, the gazetteers necessarily functioned at two scales. The first was an India-wide scale; the gazetteers covered the entirety of British India and the princely states. These volumes were authored by Hunter, and generally depicted India as a diverse but single administrative and territorial unit. This India-wide project was designed especially for general use by nonexperts, and eventually led to the publication of Hunter's monograph, *The Indian Empire: its peoples, history, and products* in 1886. The second scale was that of the Provincial Gazetteers, which were far more varied and diverse both in approach and content, and never fully succeeded in achieving the uniformity and consistency that Hunter valued so highly. The Provincial Gazetteers were overseen by the provincial governments, and the data contained within them varied dramatically, given the uneven and asymmetrical administration deployed by the British in India.

Geographical scales were not the only scales with which Hunter was concerned. Metcalf has observed the importance for the British of understanding and mobilising a particular set of narratives of time, the Indian past and the Indian future in constructing a historical justification for their rule (1994, p. 6). Unsurprisingly, therefore, history was very much a part of Hunter's remit. He wrote:

> If the history of India is ever to be anything more than a record of conquest and crime, it must be sought for among the people themselves. Valuable historical materials had been collected for the Statistical Survey; and in 1877, the Secretary of State for India decided that a wider scope should be allowed me for their use in the Imperial Gazetteer. I have done my best to give effect to that view; and it will be seen, for the first time in these volumes, that every Indian District has its own history. The true territorial unit of Indian history is, indeed, much smaller than the British District. (in Scholberg, 1970, p. v)

In short, the gazetteers were central to the idea of a 'territorial history' of India. Such a history served two purposes. The first was practical: by understanding the relationships between the Indian people and the territory they inhabited, the British could design more effective systems of governance and administration, which was especially important in a fluid frontier region characterised by a long history of invasion and conflict and populated by what the British saw as 'martial races'. But second, and as Barrow argues, such an emphasis on history would also make 'maps more appealing, authoritative, and beguiling' and that 'map makers would not only show how the present political situation was derived from what had occurred previously' (Barrow, 2004, p. 1). The notion that history is rooted in territory, and that territory itself can have a history, became a defining feature both of colonial accounts of Indian history and nationalist imaginings of a new independent India. Thus, histories of Indian territory became central to ideologies of Partition; the contested history of Indian territory quickly became grafted onto histories of the people who inhabited that territory, and the inextricable links between individual, community and territory were a central component of Indian nationalist discourse during the final decades of Empire. But such a link between territory and population was neither inevitable nor natural; in fact, the colonial discourses of identity and difference on display in the *Imperial Gazetteer of India* proved fluid and shifting even until the turn of the century.

For Hunter, the British period in India was the culminating event *not* of British history, but of *Indian history*. By placing the British within the context of Indian history, Hunter attempted to historicise British involvement in the subcontinent as a way of legitimising the 'colonial present' of the late 19th century and of the colonial Punjab as what Tan Tai Yong has termed a 'garrison state' (Tan, 2005). The gazetteers bring the story of British territorial conquest of India into the sphere of *Indian* history, a process which was formalised in 1907 when the first collection of atlas maps was included in the *Imperial Gazetteer of India*. The historical sketch maps (Figures 2.1 and 2.2), drawn by J.S. Cotton who had been Hunter's assistant on the first editions of the gazetteers, show the 'relative extent of British, Muhammadan and Hindu power' in India (Hunter et al., 1909). Shown in the customary imperial red, the historical maps provide a cartographic representation of the historical story of British rule contained in the gazetteers, which had, since Hunter's first publications in the 1880s, become the standard narrative of the development of 'India' for the British colonial government. They show British territory in the years that have become, in British historiography, the key moments in the territorial development of the Raj: '1765 (the year of the Diwani grant), in 1805 (after Lord Wellesley), in 1837 (the accession of Queen Victoria) and in 1857 (the mutiny)' (Hunter et al., 1909). They depict the shrinking territorial control of India by Indians, providing the visual-cartographic representation for narratives of both British legitimacy and authority, and of Indian claims to territory through a historical-geographical legacy of legitimate rule. We also see the gazetteer deploying the problematic term 'mutiny' to describe

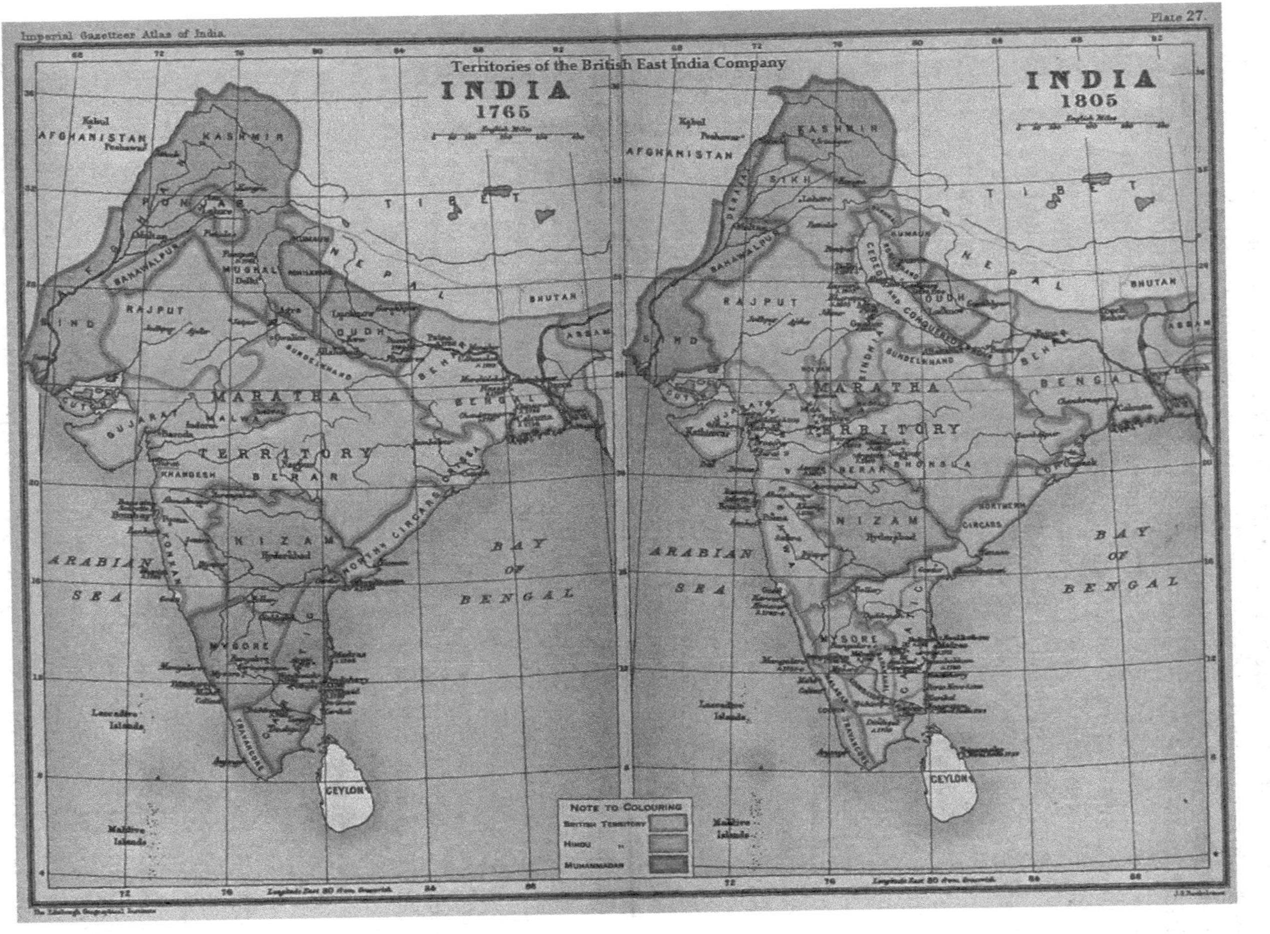

Figure 2.1 Historical maps of India, 1765 and 1805.

Source: The Imperial Gazetteer of India, vol. 26, 1909 / Isha Books / Public Domain.

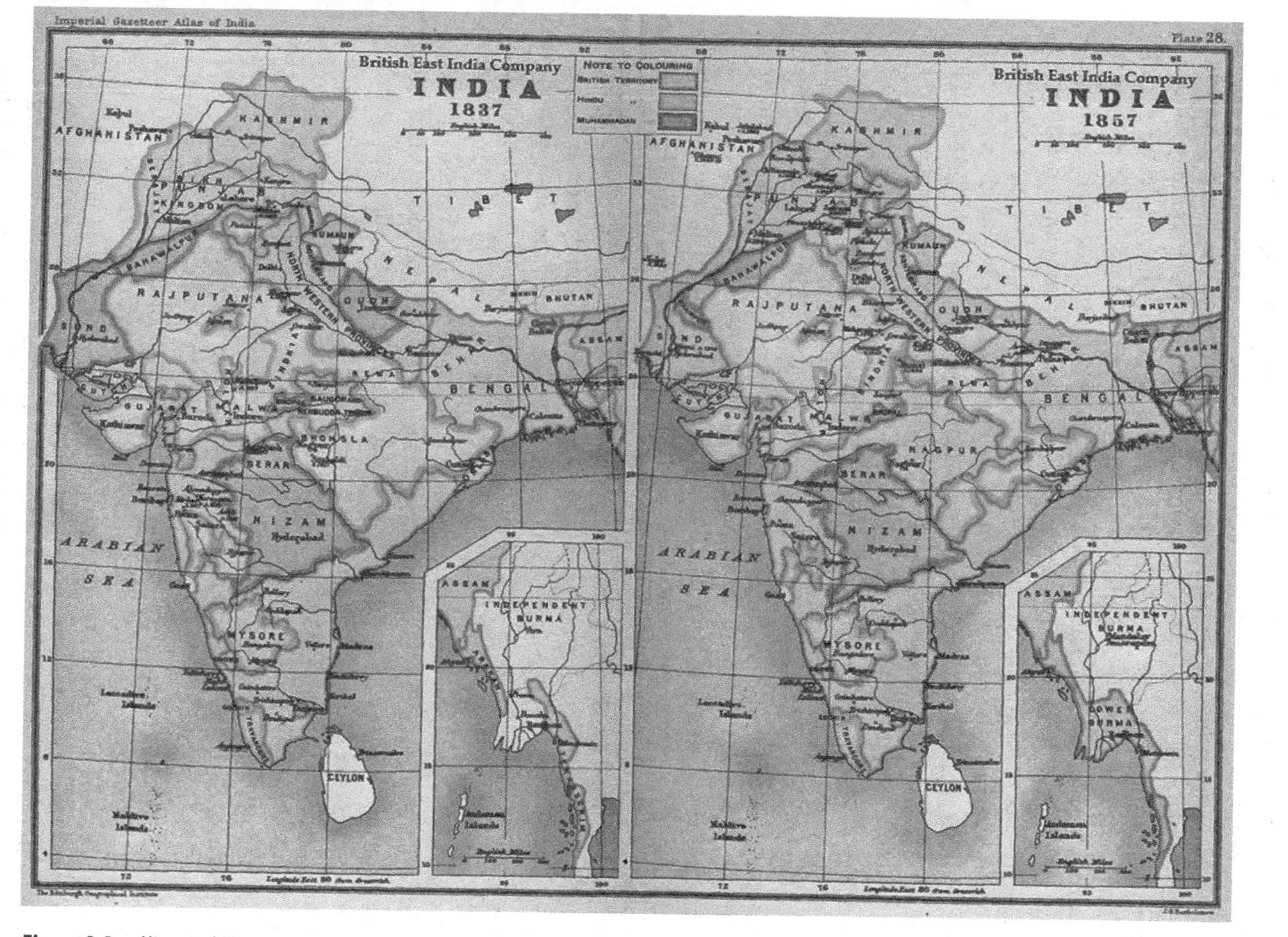

Figure 2.2 Historical Maps of India 1837 and 1857.
Source: The Imperial Gazetteer of India, vol. 26, 1909 / Isha Books / Public Domain.

the uprising by Indians in 1857, inscribing both the date and the concept into the colonial historical geography of the subcontinent.

The gazetteer project was made more complex by its multi-scale aspirations. The provincial series were to be made up of multiple volumes pertaining to each district, where information could be layered, with more specific and detailed local information organised within district and provincial volumes, while the most important and useful general information could be found in the all-India series. They were intended to refer to one another, to be used in tandem depending on the need and context. The Punjab provincial gazetteer was first produced in 1888–1889, by the British government in the Punjab, and drew heavily upon the 1881 census reports, and the Punjab Administration Report of 1882–1883. The Punjab-specific data contained within the provincial gazetteer was organised, like the all-India versions, thematically and geographically, and included extensive information on geology, natural and administrative boundaries, the railways and canal systems, the population (including extensive ethnographical data which had been collected by Denzil Ibbetson, a well-known administrator in the Punjab who had worked on the census), history, and industry in the Punjab. The history of the Punjab culminates, in the provincial gazetteer (as in the all-India gazetteers), in the arrival of the British and the instatement of a British administration in the majority of the province. Additionally, the categories of ethnic groups, understood in racial and religious terms (for example, Muslims are often not simply described as Muslims, but are also linked to regions and language groups, whilst Hindus are often denoted also by caste and livelihood), constructing a matrix of identities which were located, through historical and ethnographical study, within the fixed geographical specificities of the Punjab. Additionally, and importantly for the later analysis of the partitioning process in the Punjab, the provincial gazetteer illustrates the ways in which the Punjab had been developed, after its annexation in 1849, as a unified territorial region, connected via a complex railway system, underpinned by an agricultural economy that was growing due to new canal systems. The British attributed the particular difficulty in governing the Punjab to the province's many different racial and religious groups and stressed the effectiveness of centralised administration and the enforcement of law and order. It was this second aspect of governance in the Punjab that made the province particularly unique in the context of the rest of British India.

Hunter's method for determining the most 'original and complete' account of Indian history, like his academic contemporaries who called themselves 'Oriental scholars', relied heavily on textual sources of both British and Asian origin. Indian knowledge was recorded and transmitted in other ways, of course, including through ritual, oral transmission, music and sacred performance, but manuscripts and texts became a prime source for European scholars who wanted to understand the philosophy and history of the so-called 'Orient'. This text-based method effectively placed Indian and British histories within a single temporal 'sphere' because textual sources could be dated and therefore compared. Indian sources were, for British historiographers like Hunter, considered to be historical documents that were capable of being analysed on the same terms as British sources. He wrote that his narrative of Indian history was drawn 'from its original sources', and that he often 'had to expose old fables, or to substitute truth for long-accepted errors' (Hunter, 1886, pp. xi–xii). Obviously, such an approach denied the possibility that the political, social, and intellectual-philosophical contexts in which Indian sources were produced might differ from the contexts in which British sources were produced; more importantly, however, this approach allowed for the transformation of small-scale British historical accounts of British involvement in India into a more unified, systematic, and authoritative account of Indian history. By inserting the British into Indian history, Hunter could show that the British *belonged* in India and that they were the rightful and legitimate rulers of India. A teleological narrative of British rule in India like Hunter's lent authority to the British government and legitimised its power. British authority was facilitated by the narrative of difference and division within Indian society that was deployed by the colonial state in its bordering practices and gazetteering activity.

The *Imperial Gazetteer of India* was organised alphabetically by place name; this format was also adopted by Thornton two decades before and facilitated easy consultation and access for administrators and others who might use the volumes for practical purposes. But this format was less useful for more general or nonexpert use. Hunter therefore reorganised and condensed the information contained in the Gazetteer for his monograph, *The Indian Empire*. The book is composed of 25 chapters, the majority of which narrate Indian history chronologically, with each chapter corresponding to an epoch, with subsections devoted to racial and tribal classifications, language, literature, music, art, religious and cultural practice and political organisation. Chapter IV, for example, is titled 'The Aryans in Ancient India' and relies heavily on British Orientalist readings of Vedic texts, archaeological research and European theories regarding racial groups and migration. Chapter V is titled 'Buddhism in India (543 BC to 1000 AD)', Chapter VI 'The Greeks in India (327 to 161 BC)', Chapter VIII 'Rise of Hinduism (750 to 1520 AD)' and, eventually, Chapter XV: 'History of British Rule (1757 to 1885 AD)' (Hunter, 1886, pp. xixviii).

Rather than being organised by *geographical reference points* (like the *Imperial Gazetteer* volumes), *The Indian Empire* is organised by *temporal reference points*. This is certainly not new. Periodisation and chronology had, of course, been common in British historical narratives of Indian history, as Pandey (2012) notes, but the condensation and consolidation of many strands of Orientalist debate and research within the authoritative text of Hunter's monograph served to elide the contested and messy discursive field of Indian history. Hunter himself remarked that this process was necessary for the production of a single authoritative text on India: 'continuous condensation, although convenient to the reader, has its perils for the author. . . In attempting to reconstruct

Indian history from its original sources in the fewest possible pages, I beg Oriental scholars to believe that, although their individual views are not always set forth, they have been respectfully considered' (Hunter, 1886, p. vii). The effect of this was to create a circular authority whereby scientific language about India reinforced the accuracy of Hunter's official historical narrative, while the scientific discourses of population and territory were contextualised within a broader historical narrative of Indian progress. The census and survey data contained within the *Imperial Gazetteer* were more explicitly historicised, and the racial and religious categories presented in the *Gazetteer* became fixed to territorial units through the authoritative language of history. At the same time, the historical narrative of India was infused with the scientific language of data, bolstered by population statistics, geological, meteorological and archaeological data. Importantly, however, as Arnold observes, this science was 'politically dependent', subordinated to the imperative of good governance (2000, p. 130).

Naming and Placing 'Difference'

Hunter's historical account of India treated 'India' as a 'nation'; like most other colonial writers, Hunter understood 'India' to be an objective territorial unit which existed prior to the British arrival in the 17th century. Hunter used the term 'nation' to refer both to the territorial region the British understood as 'India' (importantly, India was not granted 'statehood', a term which implies sovereignty and self-rule), and specifically to the various racial groups Hunter observed within the Indian population. He often used the terms 'class', 'race', 'nation' and 'tribe' interchangeably, and all were distinct from the geographical unit 'continent', illustrating Gyanendra Pandey's observation that the Western perspective that 'nationalism, nation-ness, was a Western attribute, unlikely to be found or easily replicated in the East' did not develop until after 1900, when, he says, 'nationalism had emerged clearly as the discourse of the age and strong nationalist stirrings against colonialism were beginning to be felt in India' (2012, pp. 1–2). Hunter's understanding of the term 'nation' differed from later uses, which were bounded and dictated by a more specific and politico-territorial rendering of the nation-state. And yet, Hunter's use of 'nation' and 'race' indicate the types of identities the colonial government recognised as distinctive to Indian culture and society. Importantly, he did not equate religious groups with 'nations' in the earliest editions of the *Imperial Gazetteer*. Nations were racial groups, biological and hereditary groups that could be identified by language, history, occupation and geographical distribution. For the British, and for Hunter more specifically, the diversity and differentiation of the Indian population presented a challenge for effective governance. Hunter's efforts at compiling a totalising and authoritative account of India's population and territory were an attempt at producing a system of knowledge that could overcome that challenge. Colonial renderings of racial and religious differences were a central component of that agenda.

The study of religion in India had been part and parcel of British colonial knowledge production from the 18th century. The British were deeply interested in issues of religion in India, as Kenneth Jones (1989) notes, and had applied the study of religion to their development of a colonial governmentality. Scholars of South Asian religions are keenly aware of the problematic legacies of colonial articulations and applications of religion in India, but the issue of religion in India (how it functioned, how it ordered and classified Indians, and how Indian systems of justice, politics and morality were informed by religious practice and teaching) was central to British constructions of Indian society. Understanding Indian religion, it was assumed, was vital if the British were to govern India effectively (Cohn, 1996).

For Hunter, the categories of religion and race were intertwined, where each category or group could be distinguished by a unique and relatively self-contained history. He noted in *The Indian Empire* that European writers had traditionally distinguished two races or nations within India, the Hindus and the Muslims. Hunter argued that the Indian population was, in fact, made up of a 'fourfold division', comprising 'non-Aryan tribes or so-called Aboriginal tribes', the 'Aryan immigrants from the North ', the 'mixed population or Hindus' and 'the Muhammadans'. He wrote, 'The following chapters first treat of each of these four classes separately. . . These are the four elements which make up the present population. Their history, as a loosely connected whole, after they had been pounded together in the mortar of Muhammadan conquest, will next be traced' (Hunter, 1886, p. 52). Here, we see Hunter's insistence that a temporal boundary existed between these four groups, despite the fact that, when considered together, they made up the 'Indian population' as a whole; these four classes have distinct and separate pasts, a claim which implies distinct and separate biological lineages, migratory histories and religious traditions. Hunter noted, as well, the historical period in which these separate pasts converged: the Mughal Empire. Like other colonial writers, Hunter drew upon a combination of British and Indian sources to compose his historical narratives, synthesising a diverse array of colonial, Mughal, Persian and Sanskrit documents.

This recent 'pounding together' of the four classes of the Indian population had led, so the British colonial narrative went, to the need for British governance in India. Hunter wrote, 'I shall show how the British Government is trying to discharge its solemn responsibility and indicate the administrative mechanism which has knit together the discordant races of India into a great pacific Empire' (Hunter, 1886 p. 52).

In order to effectively articulate the hereditary, racial, historical, and territorial distinctions within the Indian population, Hunter deployed what had become a common technique in British constructions of the Indian past: he based his chronology and periodisation on the rise and fall of these 'four classes', these racial and religious categories Hunter identified. The movement from one epoch to another was understood to have been facilitated by flare-ups in conflict and power struggles between these groups. 'Our earliest glimpses of India disclose two races struggling for the soil' (Hunter, 1886, p. 52). When analysing the interactions between the non-Aryan and Aryan races, for example, Hunter clearly indicated the subordination of the non-Aryan tribal groups by the Aryans; 'while the bolder or more isolated of the aboriginal races have thus kept themselves apart, by far the greater portion submitted in ancient times to the Aryan invaders, and now make up the mass of the Hindus' (Hunter, 1886, p. 69). Similarly, in his discussion of the early 'Muhammadan' rulers in Chapter X, Hunter takes the time to note that from the arrival of the first Muslims in India, conflict was a feature of inter-religious relations, claiming for example that, 'The first collision between Hinduism and Islam on the Punjab frontier was an act of the Hindus' in 977 CE (Hunter, 1886, p. 272).

In this colonial teleology, the separate pasts of each category of Indian society culminated in a clearly defined set of spatial differences that mirrored the social and cultural differences which were themselves explained by these different histories. This way of conceptualising separate pasts for each category of Indian society therefore mirrored the way that both the language and maps of the gazetteers fixed these categories onto distinct and bounded units of Indian territory. The 'administrative mechanism' Hunter was referring to above, therefore relied heavily on the control and surveillance of territory and populations through the creation and maintenance of administrative units bounded by territorial borders. In Figures 2.3 and 2.4, we can see what the colonial archive believed to be the distribution of the various races and religious groups that lived in the subcontinent. These were the then-contemporary statistics, and so represented the final spatial 'outcome', so to speak, of all the history that Hunter narrates. Importantly, these maps do more to obscure than to accurately depict the diversity, complexity and mobilities of everyday life in locales across the subcontinent. They do not even show the variety and detail of the colonially constructed and partial information contained in the statistical surveys, which ran to tens of thousands of pages. The maps show population majorities, so, like the maps used in 1947 to partition India, they hide the presence of minority populations. The map of India's races (Figure 2.4) depicts categories which were, of course, rooted in European racist science and which have been debunked, but the map of India's religions (Figure 2.3) makes use of religious categories (e.g. 'Muslim' and 'Hindu') that remain recognisable as politically and socially salient in and beyond the subcontinent.

The ultimate purpose of differentiating between perceived divisions in Indian society was to create a body of knowledge that would best inform the government on the effectiveness and utility of its policies and practices. Hunter understood many of the conflicts that took place in the subcontinent before the arrival of the British in terms of power struggles over territorial control and viewed the British

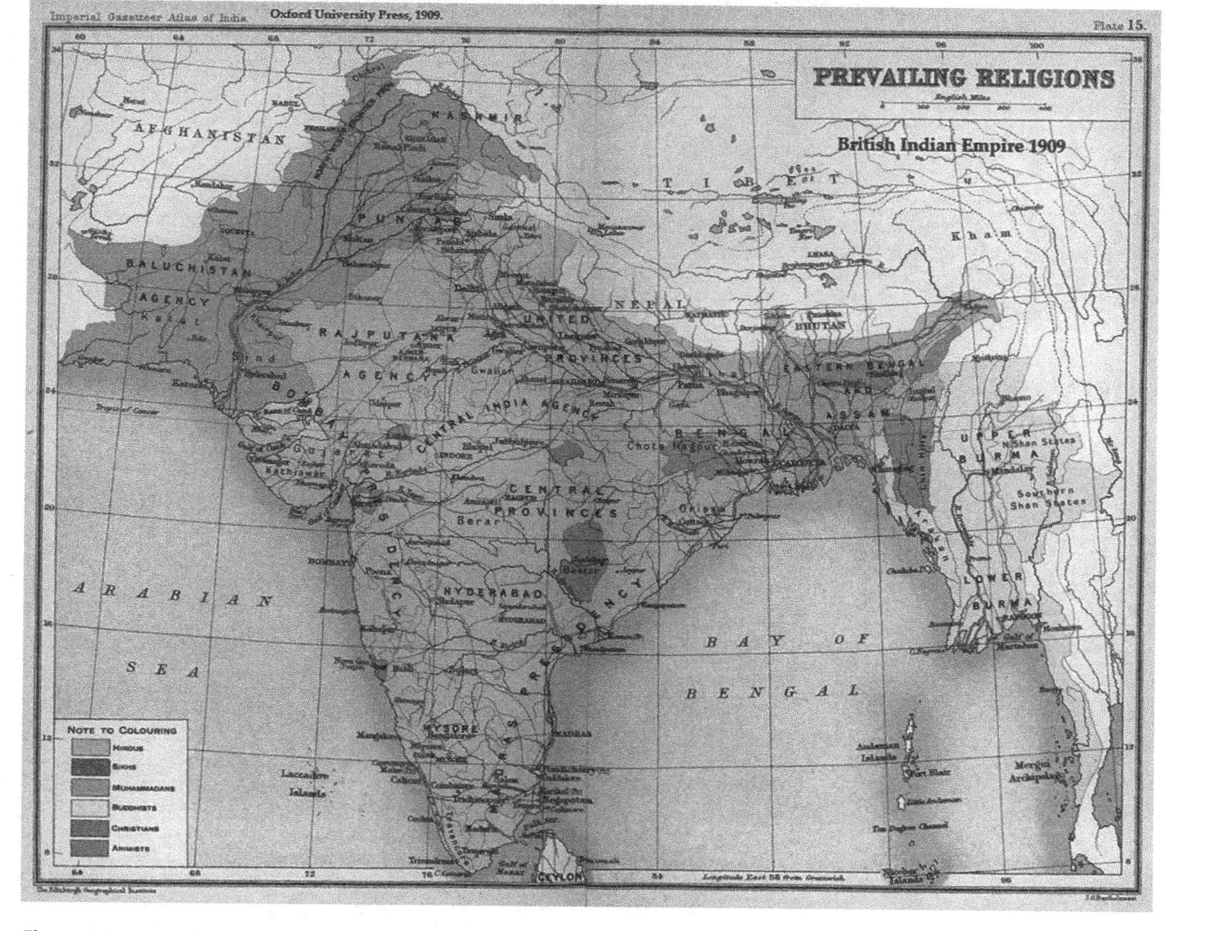

Figure 2.3 Prevailing Religions of the British Indian Empire.
Source: The Imperial Gazetteer of India, Oxford University Press, 1909/Clarendon Press/Public Domain.

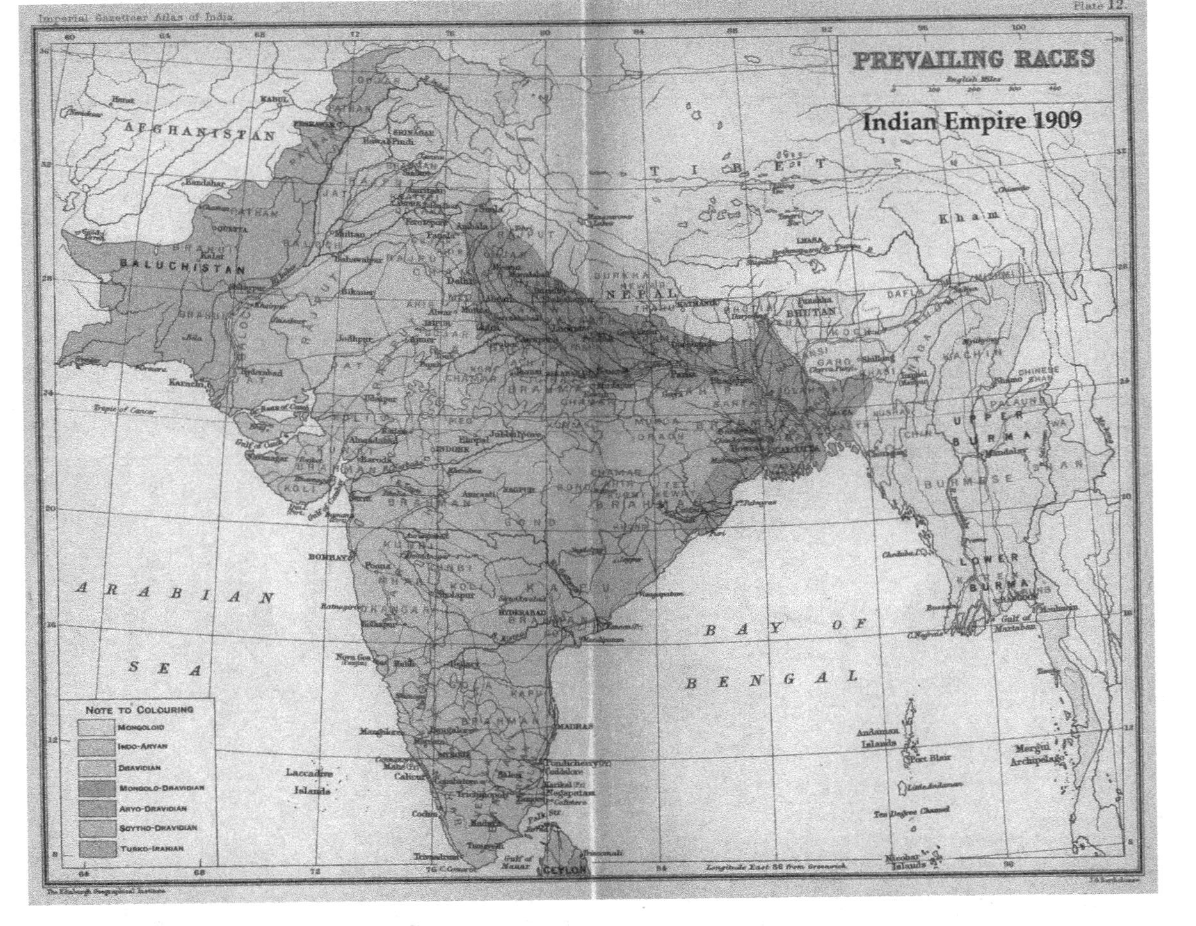

Figure 2.4 Prevailing Races of the British Indian Empire.

Source: The Imperial Gazetteer of India, vol. 26, Oxford University Press, 1909/Clarendon Press/Public Domain.

as a primarily civilising and stabilising force. For example, the transfer of territory and administration in the Punjab in the first half of the 19th century, culminating in the wars between the EEIC and the Sikhs in the 1840s, was described by Hunter in terms of the positive changes wrought by British rule. After describing the expansion and eventual wane of the Sikh kingdom in Northern India, Hunter wrote, 'At the close of the second Sikh war in the succeeding year, Dera Ghazi Khan passed, with the remainder of the Punjab Province, into the hands of our Government. Since that period, an active and vigilant administration has once more made tillage profitable and largely increased the number of inhabitants' (Hunter, 1885, p. 212). Similarly, after the same conflict, 'The District [of Dera Ismail Khan] passed quietly under British rule' (Hunter, 1885, p. 222).

The categories deployed by the British government in their quest to measure and control their Indian subjects may have shifted throughout the 19th century, but the government was consistent in their desire to represent social divisions and categories in terms of their spatial distributions. This was done not only through the census but also through the publication and dissemination of geographical literature, most systematically in the form of the *Imperial Gazetteer of India*. As noted in the introduction, Michael's work illustrates how revenue surveyors created maps which simplified on paper a more complex and messy situation on the ground, and how the social and political space shifted to match the map. Similarly, as we know, Hunter also acknowledged that India's population was made up of 'many more races and nations than Europe' (1886, p. vi). When examining the gazetteers, two key themes are evident. The first is that the British (and Hunter in particular) viewed India in a deeply contradictory way; indeed, they understood 'India' to be a unified territorial entity which they governed through a complex bureaucratic, asymmetrical, and decentralised political system. But they also saw borders and boundaries in all facets of Indian life – in its physical and human geographies Hunter regularly compared the Indian population to European populations. A closer study of these comparisons soon points to some of the ways in which European assumptions about the relationship between populations and territory work upon the colonial archive of India. The second theme is the importance of history in the legitimisation of the British narrative of the Indian population. The gazetteers provide geographical and cartographic representations of India's history, and in doing so unconsciously show the changes in imperial historiographical accounts of India over time.

The effect that this emphasis on division and diversity has on Hunter's account of Indian history is important for later justifications for the separation and division of India's population through the construction of borders and boundaries; appealing to 'History' was a powerful weapon among religious reformers, nationalists, and colonial officials alike when arguing and arbitrating India's future. More important for Hunter, however, was the importance and relevance of 'India's History' for the governance of India by the British in the 'present'. Indeed, Hunter's role as director-general of the statistical survey, and his emphasis on his methods for data collection, illustrate just how 'present' Hunter's work sought to be.

Yet despite the colonial government's desire to collate and combine its knowledge of India in the form of these gazetteers, there remained a great deal of inconsistency, variation and disorganisation in the production and dissemination of that knowledge. Colonial knowledge of the districts varied so widely that categorising these statistical and geographical reports has become a recognised challenge for archives (Scholberg, 1970). The irregularity and inconsistencies between district gazetteers illustrate the variation in British imperial and colonial knowledge and power across their Indian territories, pointing to a disjuncture between British justifications for the application of certain forms of territorial organisation and administration across India, as well as the more regionally messy and historical specific pragmatics of British colonial governmentality on the ground and in different parts of India.

Part of the logic of territorial partition can be found in this disjuncture: the measurement and application of categories and classifications of difference and division, which the British believed were necessary to govern India, played a key role in the rise of colonial and anticolonial violence. The British often employed policies and practices to more accurately and effectively describe and legislate for these racial and religious differences, when it was exactly that process that exacerbated the violence it was intended to mitigate.

Hunter's *The Indian Empire* serves as a prime example of the psychic pressure that this disjuncture placed on British rule. Like many other overviews of British power in India, Hunter's book is fixated on the ways in which the colonial government justified to itself and to its Indian subjects its economic and political exploitation of the subcontinent. Hunter based this justification on a teleological narrative of Indian history, impressing upon his readers his belief that the British had brought the most efficient, peaceful and stable government to India. Yet his book also showed that as Britain extended and consolidated its control in India, the scope and nature of this exploitation (and the putative need for it) had solicited new forms of resistance.

One of the ways that British officials and politicians in India sought to impose stability and order within British India was, of course, through bordering practices across multiple scales. Thomas Simpson (2021) has shown how peripheral border regions, namely the frontier regions, were administered and militarised in particular ways in order to ensure security and stability in more central regions. Boundary-making was, therefore, central to both colonial military and colonial politics, and was conceived by the end of the 19th century by many of its practitioners as both a science and an art that could and should be perfected. The 'ideal' border was both a reflection of and an improvement on putatively historical or cultural units that could be measured, mapped and visualised, and its effectiveness was a signifier of the broader stability and success of empire. Perhaps the most celebrated and fascinating example of the relationship between the colonising forces of geography and the military in India at the turn of the century is found in the life and work of Sir Thomas Holdich.

Thomas Holdich and the Colonial Frontier

After a successful career as a military surveyor and cartographer in the frontier regions of British India, Holdich was widely regarded as an expert not only of North Indian and Central Asian geography but of political boundary-making more generally. Holdich was a Royal Engineer, educated first at the Royal Military Academy at Woolwich and then at the School of Military Engineering. He first arrived in India in 1865 and began his long career with the Survey of India office later that year. He began his work on the Afghan Frontier in 1878, where he was assigned to the Southern Afghanistan Field Force. He served in the Afghan War, during which 'the policy of the Army towards geographical and topographical exploration' changed dramatically: 'For the first time, apparently, trained staffs of sapper surveyors, both triangulators and topographers, with Indian assistants, accompanied the advanced forces in the field to carry out scientific survey, as distinct from rapid military sketches' (Mason and Crosthwait, 1930, p. 210), which had been the main source of geographical surveys in conflict regions previously. Holdich's work (and, of course, the work of his colleagues, many of whom had been trained dually in the sciences of surveying and warfare, and many of whom were unnamed Indian experts) included both the production of professional geographical material and active military duty.

Holdich was then appointed as head of the Baluchistan Survey Party, a post he held until 1898 when he retired. He was, as the writers of his obituary were proud to point out in 1930, 'desperately anxious to leave no blank space unfilled upon the map' (Mason and Crosthwait, 1930, p. 210). The iconography of the 'blank space on the map' had fuelled much of the geographical exploration that took place during the late 19th century, particularly in parts of 'undiscovered' Africa and central Asia. Felix Driver points to Joseph Conrad's 1924 essay published in National Geographic, in which Conrad bemoans the 'irreversible closure of the epoch of open spaces, the end of an era of unashamed heroism' (Driver, 2001, p. 4).

There are echoes here of Halford Mackinder's assertion that, by the turn of the century, there were no undiscovered lands left on the map, and that this 'closed' system created a new geopolitical framework. Holdich was at the heart of this colonial project, of exploring, measuring and documenting new spaces, which could be added to the growing map of Asia for the purposes of territorial expansion and territorial administration. In fact, Holdich's writings illustrate the technical and political processes by which the British Empire extended and deepened both its territorial knowledge and its administrative control of India. Those technical and political processes of surveying, boundary-making, and bordering would become (a few decades later and with some upgrades and alterations) the same tools that the colonial and nationalist governments would use to dismantle the empire and piece together the postcolonial Indian subcontinent.

In 1884, in the midst of the so-called 'Great Game', Holdich was appointed to the Joint Anglo-Russian Boundary Commission, assigned to the task of divvying up the territory of Afghanistan between Russia and Britain (Mason and Crosthwait, 1930). He was made Superintendent of Frontier Surveys in 1892, and, in 1893, after the Durand Agreement solidified the border between British India and Afghanistan, Holdich was charged with delimiting and demarcating the line. Holdich opposed the Durand Line, primarily out of his colonial military sensibilities. Security, he believed, relied on the British Indian military's right and ability to access the frontier regions in order to police the Afghan tribes.

> He always maintained that the presence of a surveyor or topographer, who knew the ground from personal observation, among the advisers of Government, would have prevented some of the errors that were made owing to lack of geographical knowledge, and the consequent political troubles on the border, both during and after demarcation. (Mason and Crosthwait, 1930, pp. 212–213)

After retirement, Holdich was commissioned to assist in the adjudication of the Chile-Argentine boundary dispute, and in 1902, he travelled to South America, illustrating how European powers were increasingly attempting to both standardise and universalise a set of boundary-making practices that could be deployed in any geographical or political context. Over the next two decades, he published a series of books while serving on the Council of the Royal Geographical Society; he served as President of the RGS from 1917 to 1919. In 1916, he published a distilled account of his accumulated geographical wisdom in *Political Frontiers and Boundary Making.*

Political Frontiers and Boundary Making (1916) is not, for the most part, a manual for drawing borders and boundaries, or a handbook or guide for the professional statesman, in the ways that Stephen B. Jones (discussed below) saw his work on the subject two decades later. Holdich's efforts are better understood as an exemplary illustration of late colonial discourse on frontiers and boundaries, and the ways in which social and cultural boundaries were understood to be intertwined with territorial and physical boundaries in India. Holdich relies heavily on the forms of knowledge set out by Hunter in the *Imperial Gazetteers*, employing many of the observations on race and environment that Hunter included in his texts, indicating, among other things, the acceptance of Hunter's observations by colonial officials and at the RGS, which advised the British government on matters of mapping and surveying, and trained government practitioners. Holdich's narratives, like Hunter's, relay observations on varied racially and historically determined territorial relationships. Holdich's interests were, of course, different from Hunter's, and shed light on the ways in which colonial boundary-making worked on two competing and asymmetrical levels. The first was the level of Empire, which at this time encompassed, among other things, the projects that history has euphemistically termed the 'Scramble for Africa' and 'The Great Game', and which carved up, and drew new colonial lines across, Africa and Asia. The second was the level of the colony, where borders and boundaries were constructed internally for the purposes of imposing order and administration.

Holdich observed that boundary-making had become a dominant mode of international diplomacy and conflict during the final decades of the 19th century, noting, 'Never was there such an era of boundary-making in English history as during the past forty years, nor do we appear to have reached the limit yet' (1916, p. 54). Boundary disputes between empires were deeply concerned with frontier regions. It was in these areas that colonial administration was at its weakest, and where confrontation between imperial powers was most likely to manifest itself. In Holdich's framework, frontiers are spaces where Empire confronts the edges of its territorial power, and where legal boundaries and borders are not yet required. Holdich ascribes a predictable teleology to the 'frontier', writing that frontiers are 'elastic', and 'occupying an indefinite area which is gradually to be assimilated with that of the districts already brought under central control' (1916, p. 76). Holdich always advocated 'filling in the map' of expanding and deepening the store of geographical knowledge the British had used to systematically understand India and build and sustain the Raj.

Even as Holdich was advancing north into Afghanistan, he was deeply aware of the Russians advancing south through Central Asia. The Anglo-Russian Boundary Commission on which he served was tasked with delimiting and demarcating a suitable buffer region between the British and the Russians. Very little thought was given to the people who lived there, of course. The various tribal and ethnic groups who inhabited the Central Asian region were not of particular concern in this imperial battle. Holdich remarks that one of the defining features of these battles over frontier spaces in Africa and Asia was that they were fought between Europeans, rather than as colonial encounters with hostile native peoples. In line with the imperial vision of this age, Holdich believed that the British had a moral right and duty to conquer African and Asian territories and civilise their inhabitants, and thus did not flag the incongruities that existed between this imperial level of boundary-making and the colonial level.

While battles between empires over frontier regions were largely concerned with constructing external boundaries, internal boundaries within colonial domains were regularly devised and revised to better suit the smooth administration of conquered territories. These boundaries were very often constructed with what the British believed were pre-existing native principles of government and territorial organisation in mind. The tactic of 'identifying' and deploying native customs and practices was central to the colonial project.

Mahmood Mamdani (1996, 2001) has written extensively about this process in African colonial contexts. British colonial officials applied a distinct and powerful 'bordering' mindset in their analysis and internal governance of India. Hunter and Holdich both demonstrate this mindset through their articulation of social and physical boundaries. While Hunter's volume, *The Indian Empire* (1896), opens with a discussion of the physical geographical boundaries that define India, Holdich discusses the importance of the Himalayas, and the Hindu Kush range in particular, in the development of Central Asia and North India. Importantly, this 'bordering' mindset allows space for all manner of complex subsets and hierarchies. In fact, in some cases, it appears that the more complex the social boundaries, the more fascinated the British colonial observers were. Caste was one of the most convoluted and complicated sets of 'native' social practices the British observed in India, and it became one of the defining features of Indian society according to British observers and rulers. Indeed, as discussed at length in the previous section, there has been a good deal of interest in the colonial construction and rule of difference, and many of these ideas inform much of the arguments presented here. However, there has been little discussion of bordering and boundary-making practices in relation to this construction and rule of difference. Additionally, the more distinct and varied the landscapes, the more likely the British observer was to fall back on exoticised and romanticised explanations of environmental determinism and tropicality (Arnold, 2011; Driver and Martins, 2005). Colonial descriptions of the Sikhs in the Punjab, for example, relied heavily on rhetorical and aesthetic links between the 'great martial races' and the harsh dry landscape in which they lived, and are made all the more effective by contrasting descriptions of Bengalis as soft and effeminate, made weak by their easy access to water (Holdich, 1916, pp. 56–57).

Colonial modes of knowledge production facilitated the creation of a set of intertwined moral and scientific discourses that were employed for the purposes of economic exploitation and political domination in India, and how such power relied upon geographical expertise for the purpose of territorial governance. Holdich was a product and producer of such expertise. Holdich believed, like most boundary-makers, that borders could be perfected using data drawn from topographical surveys, triangulation, ethnographic and social surveys and historical inquiry. Holdich's thinking illustrates how borders act as the nexus at which physical, social and political boundaries and categories meet, and he maintained that borders must be designed with the intersections between these phenomena and various 'investigative modalities' in mind. In other words, what Holdich had in mind hinged on the synthesis of what Metcalf (1994) describes as the kinds of topographical surveys, triangulation, ethnographic and social surveys. I will return to this aspect of Holdich's work below.

This geographical-cum-military boundary-making discourse has its own integral logic, whereby a border's perceived 'success' or 'failure' is determined by the relative accuracy and skill applied to the process. Holdich prized geographical expertise as a means of boundary-making, and the significance that metropolitan and colonial politicians and administrators attached to such expertise spoke to the more general imperial ideology of science and accuracy. As emblems of both territorial sovereignty and demarcation, borders needed to be authoritative from a scientific point of view – that is, rational and unambiguous means of settling actual and possible conflicts and tensions over territorial jurisdiction and belonging. The scientific authority of borders, and the scientific standards and measures by which boundary-making purportedly works, are assumed by much Partition historiography. Much of the literature on Partition deems the boundary-making process as flawed because of its failure to match the 'reality on the ground'. The geographer Reece Jones, for example, argues that the communal and political identities upon which Partition was based (Hindu and Muslim) were, in fact, colonial constructs that had little in common with the messiness and complexities of social relationships between Hindu and Muslim Indians. He writes, 'Given the complexities on the ground in 1947, the quick and clean work of Partition perhaps seemed like the only option. Had the assumption of groups and censuses been correct, this would have been a fine choice. However, they were not and the Partition of India resulted in unmitigated disaster' (Jones, 2014, p. 286). Lucy Chester writes that British maps, designed as they were for colonial control, 'did not contain information that might have allowed the boundary commission to anticipate and prevent some of the economic and social disruptions' caused by Partition (2009, p. 7).

The implication is that boundary-making might be improved, or become less distorted or less prone to political manipulation if one devises a longer and more in-depth process of geographic data collection, and one where the data is collected for a different, noncolonial purpose. The implication is, also, that the geographical data pertinent to boundary-making is robust yet inert data that is acted on by different and competing political interests rather than as a constituent of those interests and how they are constructed. I will return to this point in greater detail below when I review the ways the 'logic' and 'illogic' of the Radcliffe Line were treated in Partition discourse. Suffice to say for now that Holdich viewed borders and the ability to make and enforce them as both a symbol of advanced civilisation and a practice of good governance. He wrote, 'Boundaries are the inevitable product of advancing civilisation' and 'the necessity for the most careful separation of spheres of national activity will continue to increase until such time as the balance of power shall be so entirely under control that it will be possible to dictate to nationalities the physical limits of their existence' (Holdich, 1916, p. 2). In other words, Holdich betrays the notion developed in this section of the chapter: namely, that boundary-making, and the geographical expertise and rhetoric of science on which it rests, are internal rather than external to imperial discourse and colonial governance.

Stephen B. Jones and the Art and Science of Boundary-Making

Three decades later, as World War II was reaching its denouement in Europe, Stephen B. Jones, a geographer at Columbia University in New York, published *Boundary-Making: A Handbook for Statesmen* (Jones, 1945). This book contains much of the same logic about borders and boundaries found in Holdich's work, albeit with some key differences. By this time, Europe had already undergone one significant process of remapping, at Versailles in 1919, where the boundaries of Central and Eastern Europe were re-mapped, and not least with recently devised methods of ethnographic cartography and the expertise of leading geographers of the day, including Isaiah Bowman (for the USA) and Emmanuel de Martonne (for France). Ethnographic mapping was designed to alleviate ethnic conflict and racially motivated violence, particularly arising from the competing claims of majority and minority ethnic groups to particular regions, by mapping the population density and spread of ethnicities across the former territories of the Austro-Hungarian Empire and thus seeking new and closer territorial fits for different ethnic groups. Excellent work on ethnographic mapping in Europe after the First and Second World Wars has animated the critical history of cartography literature (see Bowd and Clayton, 2015; Crampton, 2007; Heffernan, 2007; Palsky, 2002). Ethnographic mapping, elucidated by Holdich and his contemporaries, and developed by Jones and his contemporaries, was based on the premise that ethnic identity groups could and should be spatially segregated by territorial boundaries for the purposes of order, security and that ambiguous ideal of nationalism: the pursuit of self-determination. The European context heavily informed the theory and practice of boundary-making during the periods of decolonisation in the mid-20th century.

World War II was partly an upshot of a European territorial settlement that was bitterly resented by Germany (and Hungary), but the war yielded new questions and challenges with respect to borders, and the American military and political establishment turned once more to geographers for advice. Ethnographic maps were employed as a technique for delimiting borders designed to impose order after conflict. Gilles Palsky writes about the Paris Peace Conference and the role of ethnographic cartography in determining borders in the new eastern European states, observing, 'ethnographic maps were constantly used to estimate the possible repercussions of each boundary change on the nationalities involved' (Palsky, 2002, p. 116). Similarly, Jeremy Crampton writes about the role of eugenics and American theories of race in the same context. The Inquiry, Woodrow Wilson's task force headed by Isaiah Bowman, was charged with 'redrawing the map of post-war Europe' and 'isolating both *identity* and *territory*'. 'The ultimate goal was *racial partitioning*', which would, it was believed, 'lead to stable sovereign states' and 'yield viable and peaceful' nation-states (Crampton, 2007, p. 226). While Holdich and Jones would have argued, 30 years apart, that physical boundaries (mountain ranges or rivers, for example) were preferable for the delimitation and demarcation of boundaries, the Inquiry believed that racial homogenisation was more likely to lead to stability in Europe, and emphasised the 'self-determination' of Europe's nations.

Jones had been hired by the American Secretary of State along with a number of other academic geographers. While in Washington DC, Jones worked for Samuel Whittemore Boggs, the Head of the US Office of the Geographer (created by the Department of State) and Bowman (Clayton and Barnes, 2015; Harris, 1997). It was with Boggs' support that Jones wrote *Boundary-Making*, which, as John Donaldson and Alison Williams have recounted, went on to 'set the standard for international boundary-making in the post-war world' (2008, p. 683). The book was 'taken up by boundary specialists, for the simple reason that it provided, for the first time, a single text that detailed the core components necessary to undertake the legal and practical definition and marking of international boundaries' (2008, p. 683). Donaldson and Williams argue that Jones' book was unique in that, rather than focusing 'on categorising boundaries around the world', he 'sought to define and detail the systematic process of boundary-making' (2008, p. 684). Such a book was especially valuable, of course, for those boundary-makers who did not have specialist geographical training, and was indicative of the fact that boundary-making was becoming a function of political and legal processes. Jones observed, 'Boundary-making is a hybrid process. In a very real sense it is an earth science, yet diplomacy dominates its early and most important stages' (Jones, 1945, p. 6). In summarising his work, and in line with the way geography saw itself and debated its contribution to public life, 'Boundary-making is both an art and a science' (Jones, 1945, p. 224).

Jones' work contains detailed and what he judged to be 'useful' information on the purposes and powers of boundaries and borders. Writing in Washington during the war years, with the expectation that Europe's borders would soon be redrawn once more, but also with the sense that the colonial world would likely morph into something else, Jones' treatise provides a fascinating insight into the ways in which theories and practices of boundary-making became implicated in processes of decolonisation and post-war reconstruction. He understood the central role that borders and boundaries play in a state's ability to exercise power, and how the international state system 'naturalises' boundaries by ordering and classifying land. He wrote,

> A list of boundary functions of today would almost duplicate a list of human activities. The increasing play of government in all phases of life has made international boundaries sharp and severe barriers. Few natural obstacles restrict the movement of persons, things, and even ideas as completely as do the boundaries of some states. A boundary impinges on life in so many ways that it becomes an important part of the environment. (Jones, 1945, p. 11)

Jones also provides a decent survey of much of the relevant work on boundary-making at the time. Crucially, this work has a marked tendency to highlight the importance of two key elements – and wish-images – of boundary-making that we have already encountered with Holdich: accurate and comprehensive geographical data, and the objectivity of the geographer. But this time the art and science of boundary-making is represented as a more universal project, applicable to different places and situations, rather than a practice articulated with the travails of empire and its civilising mission. Jones writes, 'The history of many territorial disputes testifies to the need for exact information, not merely when a boundary is demarcated by also prior to its delimitation in a treaty' (1945, p. 72). He cites Charles Cheney Hyde, who wrote in the *American Journal of International Law* in 1933, 'The first condition required of maps that are to serve as evidence on points of law in their *geographical accuracy*'. Hyde continued, 'The competence of the cartographer to paint a true picture depends upon his knowledge' and 'if that is meagre, and his suppositions erroneous, he offers doubtful guidance' (Hyd, 1933, p. 311). Jones added, 'Because boundary-making is in principle a continuous process, from preliminary bargaining to ultimate administration, errors at one stage have effects at later stages. For this reason, exact information about the borderland in question should be sought as early as possible in the boundary-making process. Much of this information can best be obtained in the field, by direct investigation' (Jones, 1945, p. 54). Furthermore, 'If time and the state of mind of the conferees allow, there should be a complete regional investigation covering physical, economic, and political conditions in the borderland' (Jones, 1945, p. 72). He also highlights the importance (and also banality) of objectivity, observing, 'It perhaps goes without saying that all investigations field or office connected with boundary-making should be conducted objectively'. Hyde wrote that a geographer's 'trustworthiness as a witness must depend upon the impartiality with which he paints his picture' (Hyd, 1933, p. 314; Jones, 1945, p. 74). Jones went on to argue for the use of 'quantitative checks such as sampling, in the hands of a trained social scientist' in order to combat 'the investigator's own unconscious bias' (Jones, 1945, p. 84).

Boundary-making begins with the assertion that 'there are no intrinsically good or bad boundaries'. Jones identified four universal stages in boundary-making – allocation, delimitation, demarcation and maintenance – and argued that boundaries are made good or bad both by the 'details of delimitation and demarcation' (Jones, 1945, p. 3). Once more we encounter the assumption central to historical narratives of Partition and Partition violence, that borders are inherently neutral, and that their cultural and political meanings and functions are contingent (Chester, 2009). However, Jones also argued that the quality of a border hinges on the social and political context in which it was created and functioned. As he wrote, 'A boundary's suitability and its meaning to the peoples it limits change with changes in ideas, in methods of production, in modes of warfare, in ways of life' (Jones, 1945, p. 4). He drew on the work of the Dutch-American political scientist, Nicholas John Spykman, to define a boundary as 'not only a line of demarcation between legal systems but also a line of contact of territorial power structures' (Jones, 1945, p. 9; Spykman, 1942). In other words, there is a partial recognition on Jones's part that while geographical knowledge might be more or less accurate, the work of the boundary-maker might be more or less objective, and a border might be more or less apposite, these qualities of boundary-making and border cannot be treated as wholly external to and detached from the particular contexts in which they were mobilised and the 'lines of contact' connecting or dividing specific groups. Thus, while *boundary-making* set out to standardise boundary-making in theory, Jones admitted that it was meaningless outside of the 'lines of contact' – and of course conflict – in which it was needed and practiced. His treatise was not so much logically incongruous in this respect as fluid and pragmatic.

The four stages of boundary-making that Jones identified were adaptable as well as universal. Delimitation and demarcation are the most relevant categories and stages for this analysis, and they were often conflated and confused during the partitioning process in 1947, so I will spend a few moments outlining them here. Delimitation is the legal process whereby the boundary is negotiated and drawn on a map, and formalised through a treaty or legal agreement; whereas demarcation is the process of physically marking the boundary on the ground, of making the translation from map to earth. There is, of course, the possibility for some overlap, and Jones notes the importance of leaving some leeway during the delimitation stage in order for the demarcation process to succeed. The terms of reference for the Punjab and Bengal Boundary Commissions were at the root of the discursive confusion of 'delimitation' and 'demarcation' during the partition process. The terms stated that 'The Boundary Commission is instructed to demarcate the boundaries of the two parts of the Punjab on the basis of ascertaining the contiguous majority areas of Muslims and non-Muslims. In doing so, it will also take into account other factors' (Sadullah, 1983, p. xii).

Holdich and Jones both suggest that delimitation is characterised by legal and juridical discourse while demarcation relies far more heavily on geographical knowledge and expertise. Donaldson and Williams grasp this point, noting that, 'in spite of their deeply rooted heritage in the political geography of the first half of the 20th century, delimitation and demarcation have been rarely discussed in academic and political geography since' (2008, pp. 686–687). They go on to show how Jones' definitions of delimitation and demarcation were used by legal scholars and discuss the role of these terms in underpinning the legal process of boundary-making. In the process of partitioning India and Pakistan in 1947, however, use of the terms delimitation and demarcation often departed from Jones's construal of them. They were wrenched out of the context of his treatise and into a highly charged political and legal arena, most notably in the official terms of reference for the Punjab and Bengal Boundary Commissions, where the commissions were instructed to 'demarcate' the new boundary. The boundary commissions were, in fact, 'delimiting' the boundaries rather than demarcating them. There was not much discussion at all at the boundary commission hearings about the eventual 'demarcation' of the border, the marking of the border on the ground and how that would be done. It was hardly a priority at the time. Notwithstanding, Jones advocated making boundaries visible 'by the erection of stones, signs, beacons or monuments' (Jones, 1945, p. 210).

Jones highlighted many of the issues that appear in the Punjab Boundary Commission documents, and which remain part of historical narratives of Partition: population transfer, nationalism and self-determination and social organisation. Jones noted that boundary-making was, in many cases, unable to solve tension and conflict arising from competing nationalisms and geopolitical outlooks. 'Self-determination, literally applied [to boundary disputes], tends to be a divisive principle', he wrote, and 'If self-determination is coupled with national exclusiveness and a grim clutch on absolute sovereignty, it is not a clear guide in territorial settlements'. 'National allegiance', he said, was often 'incompatible with national territorial coherence', leading to a 'problem [that] seems insoluble in the frame of absolute national sovereignty, except by the drastic method of population transfer' (Jones, 1945, pp. 27–28). Interestingly, for Jones the problem was not the limitations posed by boundaries that led to tension and conflict; rather, nationalism, and the corresponding 'existence of overlapping territorial claims' were the cause of 'border friction' (1945, p. 13). All the more reason for surmising that Jones did not quite believe in the neutrality of borders. He recognised, however, that, while border conflict and population transfer were sometimes the unfortunate upshots of boundary-making, this form of calamity 'might be called evidence for the inadequacy of nationalism as a principle of human organisation' and be used to suggest that 'no better solution will be immediately possible' (1945, p. 44).

The Partition of India and Pakistan was indeed characterised both by conflict and violence in the newly made borderlands and by large-scale population transfer. While it is reasonable to say (and many do) that Partition was characterised by a poorly executed boundary-making process in terms of the collection and application of geographical data, it was also characterised by exactly the kinds of political and cultural differences and tensions that underpinned the type of nationalism that Jones evokes. I have no intention of arguing simply that Partition was or was not a successful or failed instance of the application of geographical knowledge because such an argument implies that applied geography had the capacity (had it been done better, for example) to alleviate the political and social conflicts which the Partition process was implemented to resolve. Rather, I argue that Jones, and Holdich before him, point to some of the ways in which questions of geography matter to understand Partition and narratives of geo-political ordering, and more specifically to understand how delimitation and demarcation of new borders at the end of World War II revolved around epistemological issues of objectivity and subjectivity, the universal and the particular, and political (legal, cultural and strategic) investment in knowing India and Pakistan geographically.

Donaldson and Williams (2008) are largely concerned with the ways in which Jones' *Boundary-Making* has been and continues to be a vital resource for boundary-makers. They argue that changing technologies in mapping have opened up the possibility of adjusting some of Jones' points, but that the epistemological framework of boundary-making work has not changed significantly since his day. Borders are still, fundamentally, about securing and ordering territory for the purpose of facilitating peaceful and normal relations between states and transactions between citizens. Ideally, they should reflect the social and environmental boundaries that already exist on the ground. Of course, it is this same bordering logic which many scholars argue is often the root of boundary problems in the first place.

Conclusion

In this chapter, I have shown how colonial logics of surveying and bordering developed in India in the late 19th century and first decades of the 20th century, with the aim of showing how they both shaped and were shaped by colonial modes of knowing and seeing the Indian population. I examined the processes by which boundaries in British India were shaped, how they shifted, and the ways in which territorial boundaries reflected and were refracted through colonial renderings of social and cultural boundaries. The chapter traced the formation of a colonial (chiefly cartographic) image of India and the Punjab that helped build and consolidate a unique project of imperial control in the region. Chapter 3 complicates the emphasis in this chapter on colonial mapping and surveying by presenting a partial historical geography of some of the Indian thought and politics from the 19th and 20th centuries that was concerned with constructing, negotiating and debating Indian identities and their relationship to space and territory. It traces the development of elite Indian professional and intellectual classes that imagined and represented the Punjab in different ways (Jones, 1989, pp. 6–7). In important respects, it was the elaborate and dynamic form of colonial administration fashioned by the British, examined in this chapter, and resisted, appropriated and challenged by Indian reformers and nationalists, which I will examine in the next chapter, that both enabled the Partition of 1947 and created the near-impossible task of doing Partition, leading to the subsequent catastrophic violence and mass migration. Taken together, these two chapters explore the range of ways that ideas and practices of territorial ownership, control and management in colonial India created the discursive framework through which Indians came to justify and agitate for autonomy, sovereignty, and the right to statehood. My wider contention is that an examination of colonial processes of boundary formation in the Punjab, and their relationship with alternative Indian spatial imaginaries and projects of counter-mapping, provide a new and important historical narrative about Partition.

These chapters together present a historical geographical framework for understanding the internal colonial and postcolonial logics of partitioned territory and stress the insidious and multiple roles played by colonial geography. I document and examine how the geographical knowledge generated by the British, both for the purposes of scientific inquiry and colonial administration, became central to the ways in which Indians began to understand themselves in relation to the territory that the British were consolidating within their empire. Such knowledge was insidious or assumed because, as Sumathi Ramaswamy (2009) notes, even the most anti-colonial applications of geographical knowledge assume the authority and legitimacy of the scientific-cartographic representation of space on the map. At the same time, the British could not monopolise understanding of this scientific-cartographic tradition, and could not prevent it from being re-deployed for anti-colonial and nationalist ends. In fact, British geographical knowledge and the spatial knowledge of the nationalist movement became articulated in complex ways. On the one hand, the former provided a foundation for the latter. On the other hand, the geographical imaginations of anti-colonialism and Indian nationalism exceeded such knowledge and foundations – outdid and evaded colonialist parameters and traditions – and reached back into prior and alternative conceptions of territory and belonging.

References

Ali, I. (2014). *The Punjab Under Imperialism, 1885–1947*. Princeton, New Jersey: Princeton University Press. https://doi.org/10.1515/9781400859580.

Appadurai, A. (1993). Number in the Colonial Imagination. In *Orientalism and the Post-colonial Predicament*, edited by C.A. Breckenridge and P. van der Veer, 314–339. Philadelphia: University of Pennsylvania Press.

Arnold, D. (2000). *Science, Technology and Medicine in Colonial India*. Cambridge: Cambridge University Press.

Arnold, D. (2011). *The Tropics and the Traveling Gaze: India, Landscape, and Science, 1800–1856*. Seattle: University of Washington Press.

Barrow, I. (2004). *Making History, Drawing Territory: British Mapping in India, c. 1756–1905*. Oxford: Oxford University Press.

Bayly, C.A. (1993). Knowing the Country: Empire and Information in India. *Modern Asian Studies* 27 (1): 3–43.

Benjamin, W. (1969). *Illuminations*. New York: Schocken Books.

Bowd, G. and Clayton, D. (2015). Emmanuel de Martonne and the Wartime Defence of Greater Romania: Circle, Set Square and Spine. *Journal of Historical Geography* 47 (January): 50–63. https://doi.org/10.1016/j.jhg.2014.10.001.

Chester, L. (2009). *Borders and Conflict in South Asia: The Radcliffe Boundary Commission and the Partition of Punjab*. Manchester: Manchester University Press.

Clayton, D. and Barnes, T. (2015). Continental European Geographers and World War II. *Journal of Historical Geography* 47 (January): 11–15. https://doi.org/10.1016/j.jhg.2014.12.003.

Cohn, B. (1996). *Colonialism and Its Forms of Knowledge: The British in India*. Princeton, New Jersey: Princeton University Press.

Crampton, J. (2007). Maps, Race and Foucault: Eugenics and Territorialization Following World War I. In *Space, Knowledge and Power: Foucault and Geography*, edited by J.W. Crampton and S. Elden, 676. Hampshire: Ashgate Publishing, Ltd.

Dirks, N.B. (2001). *Castes of Mind: Colonialism and the Making of Modern India*. Princeton, New Jersey: Princeton University Press.

Donaldson, J.W. and Williams, A.J. (2008). Delimitation and Demarcation: Analysing the Legacy of Stephen B. Jones's Boundary-Making. *Geopolitics* 13 (4): 676–700. https://doi.org/10.1080/14650040802275503.

Driver, F. (2001). *Geography Militant: Cultures of Exploration and Empire*. Oxford: Wiley.

Driver, F. and Martins, L. (2005). *Tropical Visions in an Age of Empire*. Chicago: University of Chicago Press.

Edney, M. (1997). *Mapping an Empire: The Geographical Construction of British India, 1765–1843*. Chicago: University of Chicago Press.

Foucault, M. (1977). *Discipline and Punish: The Birth of the Prison*. Translated by Sheridan. London: Allen Lane.

Gilmartin, D. (1994). Scientific Empire and Imperial Science: Colonialism and Irrigation Technology in the Indus Basin. *The Journal of Asian Studies* 53 (4): 1127–1149.

Guha, S. (2003). The Politics of Identity and Enumeration in India c. 1600–1990. *Comparative Studies in Society and History* 45 (1): 148–167. https://doi.org/10.1017/S0010417503000070.

Harris, C. (1997). Geographers in the U.S. Government in Washington, DC, During World War II. *The Professional Geographer* 49 (2): 245–256. https://doi.org/10.1111/0033-0124.00074.

Heffernan, M. (2007). *The European Geographical Imagination*. Stuttgart: Franz Steiner Verlag.

Holdich, T. (1916). *Political Frontiers and Boundary Making*. London: Macmillan and co. limited.

Hunter, W.W. (1885). *The Imperial Gazetteer of India*, 2nd ed., Vol. I. London: Trübner & co.

Hunter, W.W. (1886). *The Indian Empire : Its Peoples, History, and Products*. London: Trübner & co.

Hunter, W.W., Meyer, W.S., Cotton, J.S., et al. (1909). *Imperial Gazetteer of India*. Oxford: Clarendon Press.

Hyd, C. (1933). Maps as Evidence in International Boundary Disputes. *The American Journal of International Law* 27 (2): 311–316. https://doi.org/10.2307/2189557.

Jones, K.W. (1989). *Arya Dharm: Hindu Consciousness in 19th-Century Punjab*. Delhi: Manohar.

Jones, R. (2014). The False Premise of Partition. *Space and Polity* 18 (3): 285–300. https://doi.org/10.1080/13562576.2014.932154.

Jones, S. (1945). *Boundary-Making: A Handbook for Statesmen, Treaty Editors and Boundary Commissioners*. Washington: Carnegie Endowment for International Peace.

Kaviraj, S. (1992). The Imaginary Institution of India. In *Subaltern Studies VII*, edited by P. Chatterjee and G. Pandey, 1–39. Delhi: Oxford University Press.

Mamdani, M. (1996). *Citizen and Subject: Contemporary Africa and the Legacy of Late Colonialism*. Princeton: Princeton University Press.

Mamdani, M. (2001). Beyond Settler and Native as Political Identities: Overcoming the Political Legacy of Colonialism. *Comparative Studies in Society and History* 43 (4): 651–664.

Mason, K. and Crosthwait, H.L. (1930). Colonel Sir Thomas Hungerford Holdich, K. C. M. G., K. C. I. E., C. B. *The Geographical Journal* 75 (3): 209–217.

McDermott, R., Gordon, L.A., Embree, A.T., et al. eds. (2014). *Sources of Indian Tradition: Modern India, Pakistan, and Bangladesh*. New York: Columbia University Press.

Metcalf, T. (1994). *Ideologies of the Raj*. Cambridge: Cambridge University Press.

Michael, B. (2007). Making Territory Visible: The Revenue Surveys of Colonial South Asia. *Imago Mundi: The International Journal for the History of Cartography* 59 (1): 78–95. https://doi.org/10.1080/03085690600997852.

Palsky, G. (2002). Emmanuel de Martonne and the Ethnographical Cartography of Central Europe (1917–1920). *Imago Mundi: The International Journal for the History of Cartography* 54 (1): 111–119.

Pandey, G. (2012). *The Construction of Communalism in Colonial North India*, 3rd ed. Delhi: OUP India.

Ramaswamy, S. (2009). *The Goddess and the Nation: Mapping Mother India*. Durham, North Carolina: Duke University Press.

Rice, B.L. (1897). *Mysore a Gazetteer Compiled for Government*. Westminster: A. Constable.

Roy, K. (2011). *War, Culture and Society in Early Modern South Asia, 1740–1849*. London: Taylor & Francis.

Sadullah, M.M. (1983). *The Partition of the Punjab, 1947: A Compilation of Official Documents*. Lahore: National Documentation Centre.

Scholberg, H. (1970). *The District Gazetteers of British India: A Bibliography*. [Poststr. 4] Inter Documentation.

Simpson, T. (2021). *The Frontier in British India: Space, Science, and Power in the Nineteenth Century*. Cambridge: Cambridge University Press.

Spykman, N. (1942). Frontiers, Security, and International Organization. *Geographical Review* 32 (3): 436–447. https://doi.org/10.2307/210386.

Talbot, I. (1988). *Punjab and the Raj, 1849–1947*. Delhi: Manohar Publications.

Tan, T.Y. (2005). *The Garrison State: Military, Government and Society in Colonial Punjab, 1849–1947*. New Delhi: SAGE Publications Pvt. Ltd.

Thornton, E. (1857). *A Gazetteer of the Territories Under the Government of the East-India Company, and of the Natives States on the Continent of India*. London: W.H. Allen & Company.

Thornton, E., Wollaston, A.N. and Lethbridge, R. (1886). *A Gazetteer of the Territories Under the Government of the Viceroy of India*. London: W. H. Allen & Co.

Chapter Three
Territorialising India and Pakistan

Introduction

In the previous chapter, I argued that the geographical idea of a territorial partition has roots in colonial cartographic norms that were in regular use by the 1880s. I showed how the *Imperial Gazetteer of India* project endeavoured to construct a synthesised narrative of Indian history, geography, and population that was both complete and concise. That project was an exemplary illustration of the ways in which the British fashioned a particular model of governmentality and control in India after 1857. But while the British aimed to fashion India as a coherent territorial unit, they never truly governed India as coherently as the official colonial archive might suggest (Guha, 2011). This was partly due to the fact that the British encountered forms of Indian intellectual and political life that were not subordinated to British ways, that developed alongside and in response to the rapid social, economic and political changes wrought by colonialism and industrial capitalism, and that were instrumental in the development of nationalist ideologies and experiments in nation-building and state-making. Many of the concepts and ideas that were articulated during this period by Indian intellectuals and religious reformers became, and remain to this day, central to the discursive construction of national identities in the subcontinent, often in competing or incongruous ways. As Sugata Bose writes, 'In the late nineteenth century and the early twentieth the Indian nation was very much in the process of its own

Mapping Partition: Politics, Territory and the End of Empire in India and Pakistan, First Edition. Hannah Fitzpatrick.

making, with a variety of individuals, linguistic groups and religious communities seeking to contribute to imagining it into being' (2007, p. 131).

This chapter therefore introduces a small number of the prominent Muslim thinkers whose ideas were foundational in or who actively participated in shaping the imagined geographies of India, Pakistan, Partition and Independence in a variety of ways. These figures are already prominent in the well-established field of Indian intellectual history, as well as historiographies of Partition, but they have rarely been read in terms of their geographical thinking. Some historians like Roy Bar Sadeh and Lotte Houwink ten Cate (2021), Shaunna Rodrigues (2021) and Faisal Devji (2007, 2013, 2014) have turned their attention to the issues of scale and political identity in Indian intellectual history. Sumathi Ramaswamy (2009, 2017) has paved the way in considering how maps and globes circulated in colonial India, and how they influenced Indian intellectual and political life. Historical geography, however, has yet to engage thoroughly with these interventions in the historiography of Indian nationalism and political thought. This chapter begins to address this as-yet underdeveloped scholarly dialogue.

The chapter explores how Indian Muslim intellectuals understood the history and conditions of Indian Muslims in the 19th and 20th centuries, and more particularly how they *placed* themselves both socially and spatially within a modern colonial social space that was rendering the territory of India in cartographic and scientific terms. The chapter has three substantive sections. The first is concerned with Muslim reform and revival in the 19th century and begins by sketching some of the key prominent Indian Muslim intellectual and religious movements that emerged in the 19th century, including what is perhaps the most well-known of these movements: the institution founded at Aligarh by Sir Sayyid Ahmad Khan. This section culminates in a historical geographical interpretation of the two-nation theory, attributed to Ahmad Khan but adopted and reimagined in the 1930s and 1940s by the poet Muhammad Iqbal and Muslim League president Muhammad Ali Jinnah. The two-nation theory, which in broad terms posited that India was made up of two distinct nations, one Hindu and one Muslim, is perhaps the most famous iteration of a foundational conceptual framework in British India for understanding the historical relationship between Hindus and Muslims.

The second section moves beyond the two-nation theory to consider how issues of scale, statistics and political representation animated much of Sayyid Ahmad Khan's criticism of emerging nationalist discourse and became a core issue for Indian Muslim leaders in the 20th century. The third section engages what I term the geographical archive of Muslim nationalism, examining some of the textual and cartographic materials that demonstrate the variety of geographical imaginaries of India and Pakistan that were at work before Partition. Here I return to Iqbal and Jinnah, but I widen the scope to include Choudhary Rahmat Ali, most famous for coining the term 'Pakistan' to refer to some of India's Muslim-majority regions. I show how many of these geographical imaginaries drew upon older forms of colonial geography, both relying on and appropriating colonial cartographic representations and geographical arguments for political ends. I also show how they build upon and critique the foundations set by Sayyid Ahmad Khan and other 19th-century reform movements, in order to fashion a vision of an independent territorial state that could represent and protect India's Muslims.

For scholars of South Asian history and politics, these figures and their contributions to Indian thought and politics are likely to be well known. But for many geographers who may be less familiar with the Indian context, these materials are crucial entry points into understanding both the competing geographical imaginations of some of India's most influential Muslim thinkers and the ways that the partitioning process mobilised some of these territorial visions while sidelining others. It is difficult to find sustained engagements by these thinkers with the particular forms of geographical and spatial knowledge that are so prominent in the gazetteers and the Survey of India. This is partly a feature of the archive, which, as William Gould points out, is 'generally scattered across a range of official, semi-official and private repositories and is held in originals in multiple vernaculars' (2020, p. 15). Anthologies and collections of key primary sources, which I rely on here, reflect the history of the fields of Indian religious and intellectual history. These fields, while instrumental to my own thinking and foundational for this study, have tended not to be animated by questions of geography (prominent exceptions include Faisal Devji (2014) and Sumathi Ramaswamy (2002, 2009)), nor have geographers tended to do the asking.

This might also reflect the fact that Indian intellectuals were framing their questions about India's geography in different ways than those of colonial administrators and military geographers. Yet finding traces of geographical thinking (broadly construed) in these sources is both challenging and necessary because many of the inherent tensions and contradictions in the dominant competing nationalist agendas by the 1940s were geographical in nature. Indian thinkers and their interlocuters were engaging with concepts and ideas relating to boundaries, limits and space, whether or not they explicitly framed them in terms of European geography. Often reformers were grappling with India's place in a global or proto-transnational geography that included the British Empire but also transcended it. I argue, however, that traces of Indian geographical thinking in this period are often embedded within the competing narratives and interpretations of India's past that are represented in these texts. How can such a re-examination help us understand processes of boundary-making, or *partitioning*, in colonial India, and in particular the process of partitioning that took place in July and August 1947?

To engage with this question, I recast the broad and relatively well-known story of the two-nation theory, the emergence of Pakistan as an idea and its incorporation into Muslim League politics in historical-geographical terms, in part by calling into question the ways that historiographical work on these ideas relies on the relative ontological stability of geographical practices and projects, including cartography and mapping, territoriality and boundary-making. Following Uday Singh Mehta (1999), Bayly (2011) and others, I frame this historical geography in terms of Indian engagement with British liberalism, particularly with regard to the founding of educational institutions. It is worth noting that Indian thinkers engaged not only with European liberal philosophy but also with more radical European and international political and intellectual movements, including communism (Raza, 2020). However, I focus on liberalism here because the intellectuals and political leaders I examine (and who had significant roles in shaping the eventual boundary-making process) were in close dialogue with European liberal philosophy. Throughout the chapter, I tease out some of the ways in which historical inquiry, both in the 19th century and in the first half of the 20th, assumes certain norms and truths about space and territory, both within India and in the wider imperial and global context.

The individuals I discuss below have been put to work in a variety of contradictory and complimentary ways by nationalists, academics, activists and others. My purpose is not to interrogate the internal debates and tensions among Indian Muslim intellectuals during the 19th century, or even the nationalist debates among Indian Muslims in the 20th century; nor is it to evaluate the truth claims of contemporary historiographical projects on these movements. Rather, I hope to examine some of the ways that Indians conceptualised and imagined the spatiality and territory of the subcontinent in ways that challenged or disrupted colonial imaginaries of those same spaces.

The book is concerned especially with the Muslim League, and on how colonial geographies of religious or communal difference constructed particular ways of territorialising Indian Muslims. Therefore, I focus less here on Hindu intellectual and reform movements than I do on the work of Muslim thinkers. Ultimately, I argue here that many of the Muslim thinkers I consider were often imagining ways of overcoming the territorial bounding of identity that was so central to the colonial project, by emphasising mobility, the complexity of subjectivity and identity, and the importance of scale, in order to protect diversity and difference in India without undermining its coherence as a political unit and Indians as a collective who were worthy of justice and good governance. This often was articulated differently from how Hindu intellectuals imagined India, especially India's historical geographies, which often overlapped with colonial historical narratives that constructed a contrast between Indian Muslim migration into the subcontinent and Indian Hindu stasis in the subcontinent. Many Hindu thinkers, including many associated with the prominent Arya Samaj school for example, often articulated their claims for political representation for Indians through a territorial claim based on continuity or stasis (for a thorough examination of the Arya Samaj, see Jones (1989)).

I argue that these complicated ways of understanding Indian Muslims' place in modern India did not always map intuitively onto European conceptions of the nation-state that were developing concurrently in the 19th century, and were therefore often difficult to territorialise. None were able to entirely overcome the tensions in such work (tensions which eventually led to Partition in 1947, as well as the inability of Partition to actually resolve those tensions), and contemporary debates around who can be a member of the nation-state remain at the heart of political and social life in the subcontinent today.

Indian Muslim Reform and Revival in the 19th Century

This section introduces the wider 19th-century social and political milieu within which Indian Muslim intellectuals were thinking, writing and debating, to make clear how their interventions were part of a wider context of reform and revival in India, as well as a distinct and important set of debates that included considerations of geography, spatiality and territory. Religious thought and practice in 19th-century India were refracted through the changing administrative and territorial context of India, which I examined in Chapter 2. Indian reformist and revivalist movements are therefore often studied in the context of the changes wrought by colonial modernity and challenges to colonialism through the development of Indian nationalism. Here I begin with the assumption that reformist thought did indeed pave the way for the rise of nationalist sentiments and activity in the 20th century, although it is important to remember that this rise was uneven, asymmetrical and regionally diverse.

Reformism and revival in India have often been characterised as a set of Indian engagements with the colonial 'modernity' that the British sought to fashion in India – in particular, as responses to the intertwined processes of Westernisation and Christianisation, the latter beginning in earnest after the government lifted its ban on missionaries in India following the East India Company's charter renewal of 1831. The arrival of missionaries in India inspired both reactionary and radical forms of religious and social reforms, all of which led to complex and tense debates within religious groups and among Indians more generally (Cox, 2002; Hay, 1988; Jones, 1989, 1992, 2006). Kenneth W. Jones (1989) argues in one of his early books that the new forms of debate, publication, and proselytising allowed for the growth of new articulations of identity and community. Most reformist movements relied heavily on new understandings and articulations of the past, of India's history, in an attempt to challenge the power of imperial and Christian discourses, but also to create new relationships between Indians to revise, revitalise and strengthen Indian traditions within the context of British India, which were often perceived as being under attack.

Indian Muslim thinkers were often engaged with thinking about Indian Muslims in this context of a rapidly changing global political and economic system, and grappling with the social and economic implications of the geopolitical upheaval associated with the transition of power away from the Mughals to the British colonial government (see Devji, 2014). They, like their Hindu contemporaries, responded to the work of Christian missionaries in their communities, and they were thinking through the material implications of colonial rule for India's Muslims, which had generally been characterised by worsening social and economic conditions. Colonial European attitudes towards colonised Muslim subjects, both in India and in other colonies, were different from attitudes towards colonised Hindu and Sikh subjects, and by the late 19th century, theories of 'Muslim exceptionalism' were prominent in colonial discourse and colonial administration (Aydin, 2017).

William Gould observes that, among Muslim movements in north India in the early to mid-19th century, 'there were common trends around the removal of erroneous innovations, purification and the simplification of ritual. More differences between Muslims in these debates came about in the contrast between intellectualism and activism' (2020, p. 5). Islamic practice in the subcontinent prior to the arrival of the British was diverse and varied. Often scholars of intellectual history and religion make a broad distinction between sufism and the 'ulama, where the former is associated with mysticism or asceticism and the latter is associated with scholarship, textual and scriptural knowledge and expertise in the Law of Islam. Both forms of Islam, with additional regional, historical and traditional distinctions, existed in India when the British arrived. Indeed, at this time, the distinctions between sufis and 'ulama were not so clear cut, and individual Muslim leaders or teachers could embody both forms of religious practice. On the other hand, in the context of powerful Muslim states, sufis and 'ulama also found themselves vying for political patronage and influence. The Mughal court, for example, had an affiliation with the sufi Chishti order, which extended back to the medieval period, while at the same time, the Mughal state structure provided opportunities for the development of the 'ulama in the form of bureaucratic power and involvement with the running of the state (Metcalf, 1982). Ultimately, Indian Muslim religious leaders, whether sufi or 'ulama, were often entangled with the Mughal state.

Despite the colonial government's ideology that assumed a lack of mobility, or stasis, on the part of its colonised subjects, the historical geography of Islam that came to dominate the narratives produced by Indian Muslim thinkers in the 19th century emphasised mobility and movement. The educated elite were well-travelled, and as Barbara Metcalf (1982) notes, travel between hubs of Islamic learning was central to the training of the 'ulama in India and elsewhere. Indian Muslim religious leaders were part of a network of elite Islamic schools and organisations, while at the same time, they developed philosophical schools that were distinctly South Asian which were themselves sites of travel for Muslim scholars from outside the subcontinent. Indeed, prominent Indian Muslim politicians and thinkers from the 20th century, including Muhammad Iqbal and Maulana Abul Kalam Azad, were shaped in large part by their mobile lives (Majeed, 2007; Rodrigues, 2021).

Sufism in northern India also had a particular historical geography, which was characterised by networks of local shrines that became hubs for worship and pilgrimage. In Punjab and Sind, particularly in the rural countryside, Muslim religious practice was dominated by the *pirs*, or hereditary religious leaders, who were associated with these particular sacred localities, which were medieval tombs or shrines of sufi saints. Sarah Ansari (1992) analyses the relationship between the sufi *pirs* in Sind and the British colonial government, noting the significance of the *pirs* in the social and cultural arenas of rural Sind. *Pirs* 'have acted as "hinges" or "mediators" between God and man'. . . 'To an overwhelmingly unlettered following, they came to symbolise what it meant in practice to be a Muslim' (Ansari, 1992, p. 9). Similarly in Punjab, as David Gilmartin points out, 'it was these local centres. . .which provided the focus for Islamic organisation in most of rural western Punjab, and it was to these centres that the population looked to for religious leadership' (1979, p. 486).

Historians tend to frame South Asian Islamic movements of reform and revival not simply as a response to the arrival of the East India Company and then the British colonial state but also in terms of their nuanced and varied responses to the decline of Mughal power in the subcontinent and beyond beginning in the 18th century (Ansari, 1992; Devji, 2014; Gilmartin, 1979; Metcalf, 1982). For the sufi pirs in Punjab and Sind, this decline meant that they were able to consolidate their influence at the regional scale, and revitalisation movements enhanced both the spiritual and political authority of the hereditary sufi leaders. In Punjab, 'the eighteenth century saw the revitalisation of the Nizamiyyah branch of the Chishti Sufi order as a major response to the decay of the imperial institutions' (Metcalf, 1982, p. 27). David Gilmartin implies a particular geographical sensibility that underpinned this revival of Chishti Sufism, which relied on the missionary activities of followers to spread its teachings, and which resulted in the establishment of 'a network of extremely influential Chishti shrines' which 'became major new centres of religious authority' (1979, p. 490). Gilmartin emphasises that these shrines retained and 'continued political structure of rural Punjab society, where power was diffused among a large number of rural, often tribally based leaders' (1979, p. 491). This was in contrast to the movements of the 'ulama, which developed out of the urban centres where Mughal power had been strongest and most centralised.

Among the 'ulama, prominent 19th-century movements developed following the earlier influence of the 18th-century writer and teacher, Shah Wali 'Ullah (1703–1762), whose work attempted 'to develop new forms of organisation to produce an independent class of 'ulama which could set religious standards for the community' (Gilmartin, 1979, p. 490). Responding not just to the arrival of the British but also to the rise of the Sikh and Maratha polities, he sought to carve out a clear role for the 'ulama as advisors to political leadership and as teachers and guides to the Muslim community (Metcalf, 1982). His movement inspired a number of distinctive offshoots, some of which fervently rejected sufi forms of worship and the religious authority of the *pirs*, while others sought to reform sufi practices, and incorporated them into their reformist agendas. Prominent among these was the movement led by the revivalist Sayyid Ahmad Barelwi, who studied with the heirs of Shah Wali 'Ullah, and who eventually cultivated a following at Rai Bareilly. A charismatic leader who emphasised religious practice over intellectual interpretations of text, Sayyid Ahmad developed a vision of a renewed and revitalised Muslim state, carved out of the frontier region. He was most famous for his declaration of *jihad* against the Sikh ruler Ranjit Singh, and he eventually died in 1831 fighting the Sikh empire in Balakot. After his death, the movements led by 'ulama moved away from warfare, and towards other forms of negotiation with the state.

A different type of reform arose around Indian Muslim institutions of education. Prominent among these was the madrasa founded at Deoband in 1867. Barbara Metcalf, whose work on the Deoband school in the 1970s was foundational, has called it 'a model for religious education in modern India' and notes that the school represents 'an important incipient trend toward a formal bureaucratisation of the ulama and their institutions' (1978, p. 111). They sought both autonomy and distance from the state, and 'to create a body of religious leaders able to serve the daily legal and spiritual needs of their fellow Muslims apart from government ties' (1978, p. 112).

A smaller revivalist movement, very much not affiliated with the 'ulama, and less prominent in the historiographical literature, developed in Punjab around the figure of Mirza Ghulam Ahmad, which later came to be called the Ahmadiyya movement. I introduce them here because of the role they played during Partition. Mirza Ghulam Ahmad was born in 1835 in the town of Qadian in Punjab. Officially established in 1889, the Ahmadiyya movement shares most of its core beliefs with other Sunni schools of Islamic thought, but differs in one fundamental way. While most Muslims generally agree that Muhammad was the final prophet, the Ahmadis understand Mirza Ghulam Ahmad to have been a prophet and a messianic figure. This belief rests upon a particular revisionist interpretation of the relevant verse in the Quran. Ali Usman Qasmi argues that this too should be contextualised by the end of the Mughal empire: 'The decline in Muslim political authority and concomitant ascendancy of the Christian and colonialist West from the eighteenth century onwards was the commencement of one such phase in Muslim history where anticipation of and yearning for messianic figures increased' (2014, p. 36).

Ghulam Ahmad's position on the finality of the prophethood of Muhammad angered many of the 'ulama, in his own lifetime and after. The ways in which debates unfolded between the different sects and schools were characterised by differences in interpretation and in practice, but all of the reformist and revivalist movements were redefining what it meant to be a Muslim, both in relation to other Muslim organisations and in relation to the wider debates playing out across religious boundaries. However, Ali Usman Qasmi (2014) notes that despite the many distinctions, the importance of the Prophet Muhammad in articulating both a moral model for Muslims and in delimiting a clear boundary between Muslims and non-Muslims was central to Muslim reform in British India. 'The increasingly charged atmosphere. . . made Muslims ever more conscious about safeguarding the image and honor of their Prophet' (2014, p. 40). Mirza Ghulam Ahmad preached a renunciation of militant jihad, as well as support for the British colonial government. However, some of the movements, like Sayyid Ahmad Barelwi's, were opposed to the British presence in India, and expressed anti-colonial and anti-British positions and teachings. Mushirul Hasan notes that many of them 'were grounded in the belief that the Islamic community enjoyed an autonomy of its own requiring no "external" mediation. Thus Islam in India had to be purged of "Hindu" accretions, and its followers were to be equipped to combat the pernicious influence of the West' (Hasan, 1993, p. 6).

In practice, this often meant rejecting both Western scientific learning and practices and beliefs that were deemed too 'Hindu'. However, Sayyid Ahmad Khan (1817–1898), arguably the most famous of the 19th-century Indian Muslim intellectual and religious reformers, did not share this view. Hasan writes of him: 'In his generation, he was among the few to read the writing on the wall, recognise the presence of a powerful colonial force in the subcontinent, and, for that reason, laid stress on. . . innovation, reforms, and change' (Hasan, 1993, p. 88). Born in 1817, his formative experiences as a colonial subject in India led him to believe that British rule, if reformed, was the most desirable political system for India. He condemned the 1857 uprising and, while he chastised his fellow Indians for their use of violence, he ultimately believed that the British were responsible for reforming their government in order to avoid another rebellion. Ahmad Khan was famous (and encountered a great deal of opposition) in his own lifetime for his philosophies of syncretism, bringing science and Islam into the same intellectual sphere, and for his work as an education reformer. In 1866, he founded the Scientific Society at Aligarh.

He was deeply committed to his belief in the value of formal education for Indian Muslim social uplift, establishing the Mohammedan Anglo-Oriental College (MAO) in Aligarh in 1877, an institution which eventually sat at the heart of Muslim intellectual life in India, and which remains a significant and prestigious institution of higher education. Modelled on the University of London and the residential college system of Oxbridge, the college at Aligarh was founded with the mission of educating Indians in both the religious and philosophical traditions of India and in the arts and sciences of Western learning. He believed that Western learning was complimentary to Islamic philosophy and practice, and that Indian Muslims should not only be encouraged to pursue an education in the Western arts and sciences but that they were in fact obligated by their faith to do so. Hasan (1993) notes that 'his message embodied, in large part, the *ijma* (consensus) of succeeding Muslim generations in the subcontinent. His interpretation of Islam was to become part of the furnishing of the mind of educated Muslims' (1993, pp. 88–89).

During the last two decades of the 19th century, much of this work was done through the promotion and expansion of an educational curriculum which could train Indian children for lives as productive and effective Indian subjects whose work would benefit the 'nation'. Manu Goswami examines what she calls 'colonial pedagogical consolidation', during the second half of the 19th century (2004, p. 132). She analyses the relationship between the dissemination of globes and British geographical training in schools in India and the rise in nationalist articulations of India as a geopolitical territory that could be located on the globe. In particular, she argues that nationalists in India had, by 1880, absorbed the colonial construction of India as a place that could be examined empirically from a distance (Goswami, 2004, p. 132). Ramaswamy similarly argues that by the turn of the century, Indian nationalists drew heavily on British geographical training to construct their territorial (and cartographic) depictions of India, to 'visualise India's geo-body', as the title of her 2002 paper suggests (Ramaswamy, 2002, 2009). In *Terrestrial Lessons* (2017), Ramaswamy explores how the globe functioned as a pedagogical tool and mode of reimagining India's place and position in both physical and cosmological terms. Subho Basu meanwhile examines how the Bengali Hindu literati drew on European geographical knowledge about race and civilisational hierarchies, which were reproduced in Indian textbooks in the 19th century, to 'adjudicate claims of different Indian communities about their homeland that would later inform the interconnections between nationality and territorial claims of the putative national state' (2010, p. 58). These scholars show how European geography was incorporated in diverse and uneven ways, but that it was not directly challenged by Indian thinkers in a systematic way. Indeed, it is difficult to find *direct* challenges to European scientific geographical knowledge beyond the Hindu cosmological renderings of space and geography that Indian writers of geography were either refuting, reworking or synthesising. The intellectuals I explored here were thinking within and around European geographical explanations of the physical and social landscape of India and the world.

The Two-Nation Theory

1940s and post-1947 Pakistani nationalist narratives often cite the two-nation theory as a scholarly justification for Partition and as the intellectual basis for the creation of an independent nation-state for Indian Muslims after Independence. Historians of Partition and scholars of Indian religious and intellectual life have often examined the intellectual 'life', so to speak, of the concept, which holds that India was made up of 'two nations', the Hindus and the Muslims, and that these two 'nations' were distinct from each other in history, religious practice, culture and language. The idea itself has been mobilised to justify varied and opposing views on the legacy and legitimacy of Partition. I am interested in the ways in which Sayyid Ahmad Khan drew upon and challenged colonial discourses of territory and difference that were in the process of shaping and consolidating modern India in his time. I argue that, ultimately, Sayyid Ahmad Khan attempted to transcend the essentialist territorialisation and historicisation that was part and parcel of both the colonial and to a significant extent the Hindu geographical imagination of India, which constructed a long historiography that often emphasised the rootedness of Indians within the territory of India.

Despite his prominent role in the historiography, Ahmad Khan's intervention was not the first time that the history of India was articulated using religious or so-called communal difference as a mode of explaining the social and political world of the subcontinent. Manan Ahmed Asif (2020) shows how historical narratives of religious difference captured the imaginations of East India Company officials in the 18th century. Similarly, the two-nation theory was one iteration of plural and competing intellectual ideas that grew out of this period, but it has taken on both an intellectual and symbolic significance in intellectual histories of nationalism in the subcontinent.

In 1883, Ahmad Khan delivered a speech in Patna, in which he famously said, 'My friends, I have repeatedly said and say it again that India is like a bride which has got two beautiful and lustrous eyes – Hindus and Mussulmans. If they quarrel against each other, that beautiful bride will become ugly and if one destroys the other, she will lose one eye' (in McDermott et al., 2014, pp. 151–152). He invoked an idea of difference that was in some ways similar to Hunter's use of the term 'nation' in the gazetteers, used to describe the various ethnic, linguistic and religious groups the British were cataloguing in their quest to compile a complete knowledge of India, but Ahmad Khan's was a more nuanced interpretation than Hunter's of the territoriality of those differences. Ahmad Khan's 'nations' of India were religious communities which were distinctive not only in the bounded and observable differences in religious practice but also by their varied histories of mobility and home-making in India. Importantly, however, he insisted that accounts of Indian history could and should make space to incorporate the multiplicity of pasts among India's people. By doing so, he could then argue that the India of his time was made up of two nations that actually shared a more fundamental historical-territorial relationship with India: one in which all Indians were united by a past and present shaped by mobility and migration. Ahmad Khan said,

> The Hindus forgot the country from which they had come; they could not remember their migration from one land to another and came to consider India as their homeland, believing that their country lies between the Himalayas and the Vindhyachal. Hundreds of years have lapsed since we, in our turn, left the lands of our origin. We remember neither the climate nor the natural beauty of those lands, neither the freshness of the harvests nor the deliciousness of the fruits, nor even do we remember the blessings of the holy deserts. We also come to consider India as our homeland and we settled down here like the earlier immigrants.-(1883, in McDermott et al., 2014, p. 151)

Parallel histories, parallel migration, and centuries of living in India were more important criteria for thinking of India as a 'homeland' for Hindus and Muslims, according to Sayyid Ahmad Khan, than the religious differences so often cited as the source of strife between the two communities.

> Thus, India is the home of both of us. We both breathe the air of India and take the water of the holy Ganges and the Jamuna. We both consume the products of the Indian soil. We are living and dying together. By living so long in India, the blood of both have changed. The colour of both have become similar. The faces of both, having changed, have become similar. . . both of us, on the basis of being common inhabitants of India, actually constitute one nation. (1883, in McDermott et al., 2014, p. 151)

Sayyid Ahmad Khan's articulation of the two-nation theory both reflects and challenges the discourses of 19th-century colonial knowledge production. On the one hand, his 'religious communities' chimed with Hunter's categories for measuring population. Like Hunter, he appealed to a linear chronology of history, but, unlike Hunter, he did so to justify his belief that all Hindus and Muslims had equal claims to both the affective ideal and the physical resources of India. But he challenged colonial imagery of social and political boundaries as markers and measures of division within Indian society, arguing that mutual cooperation and unity were more important for the prosperity and development of India than the imposition and enforcement of social boundaries between groups dictated by identity categories.

But we can further nuance our understanding of the context in which Ahmad Khan was working. He was part of a community of Indian Muslim thinkers who found themselves grappling with geographical deterministic arguments about Muslim exceptionalism, put forward by both colonial writers like Hunter and a number of Hindu writers, that implicitly, and sometimes explicitly, rendered Muslims as outsiders within India. Manan Ahmed Asif (2020) has recently outlined how British colonial history-writing projects in the subcontinent reflected and reproduced established white Christian European attitudes towards Muslims and Islam, which influenced the ways in which Orientalist scholars, colonial officials and British politicians understood the religious differences they perceived among their Indian subjects. British scholars of Indian history were often interested in the so-called 'origins' of India's populations. They drew a broad distinction between 'Muslim India', which had its so-called 'origins' outside of the subcontinent, in what was becoming known as the Middle East, and 'Hindu India', which was often rooted in Vedic interpretations of the origins of Indians. One consequence of this colonial historical project, often implied in the literature but not spelled out, was the construction of two distinct *historical* geographies of India: one Muslim, one Hindu.

This temporal distinction between pre-Muslim and post-Muslim India served a number of purposes. First, it reflected a longer trend in British colonial history-writing about India, which marked the Mughal Empire as a distinctive period in Indian history and which was characterised by foreign conquest of a pre-existing society and population. Ahmed Asif (2020) argues that histories of the subcontinent were in fact produced by scholars in the Mughal courts, and that these histories were often discredited or poorly interpreted by early British interlocutors, who themselves were working within an older European tradition that framed Muslim thought and politics in terms that emphasised violence and conquest.

Second, the temporal boundary of pre- and post-Muslim India signifies an imagined definitive and unambiguous point on the timeline of India's past when Muslims were not within the territorial limits of India. The historical record, as Ahmed Asif (2020) makes clear, is not so neat and tidy, and the development and consolidation of Muslim polities in the subcontinent was gradual, taking place over centuries. While temporal connections make up the bulk of historiographical writing on Indian nationalism, spatial analysis is also important for understanding the Indian colonial world of the late 19th century and early 20th century that gave rise to a new and interconnected Indian sphere. As discourses of territorial contiguity and political sovereignty became increasingly entangled outside Europe during the 18th and 19th centuries, this careful historical territorial othering of India's Muslim population sharpened the imagined spatial and territorial differences between Hindus and Muslims in India.

The effects of this colonial historiography can be seen even in more contemporary critical scholarship on Indian colonial history. For example, both Manu Goswami (2004) and Sumathi Ramaswamy (2009), who have produced some of the most rigorous scholarships on Indian history to date, focus on the unity, both in territory and identity, which underpinned Indian (especially Congress) demands for self-rule and sovereignty. Ramaswamy (2009) in particular hones in on what was, ultimately, a Hindu-inflected artistic and cartographic tradition of depicting the Goddess in map form (even though, importantly, some Muslims have also contributed to this tradition, and Ramaswamy acknowledges this regularly in her work). One unintentional effect of these careful genealogies, while convincing and rigorous, is to downplay the concurrent Indian Muslim articulations of India which posited alternatives to this narrative of territorial, temporal and religious unity. The two-nation theory was one such alternative. Where Hunter constructed a story of diverging pasts, distinguished between ancient Aryans as stationary and rooted and Mughals as transitory and mobile, Sayyid Ahmad Khan flipped this colonial narrative of transience on its head, to narrate a story of universal mobility and migration culminating in a present where territory is equally occupied and shared. He challenged the significance of primordial or timeless spatial fixedness in justifying territorial claims in the then-present moment of 19th century India, instead framing human history in terms of periods of movement and circulation.

After Sayyid Ahmad Khan, the two-nation theory was refashioned by Muhammad Iqbal, the philosopher-poet who became, posthumously, the national poet of Pakistan. Iqbal was born in 1877 in Sialkot, Punjab. In 1905, he travelled to Europe, where he attended Cambridge University and became a lawyer, before traveling to Germany where he completed a PhD. He wrote prolifically in Urdu and Persian, and while he wrote scholarly work in multiple fields, he is most famous for his poetry (see Majeed, 2007, 2009). He was also a prominent member of the Muslim League, in which he held multiple positions. In 1930, Iqbal addressed the general meeting of the All India Muslim League in a now-famous speech that has been interpreted and debated by Pakistani and Indian commentators concerned with nation-building and identity construction in the subcontinent, laying out his proposals for an independent India. In it, he argued for a decentralised federal system of Indian states based on religious identity. Iqbal believed that Indian Muslims were members of two communities: *a religious community*, the Muslim *umma,* which included within it many different ethnic and cultural groups, and the Indian community, the geographically unified but culturally diverse unit of the subcontinent. The *umma*, for Iqbal, was not subject to borders; indeed, one of the distinctive features of the Muslim *umma* was that it could transcend geographical and political boundaries, and could not be contained within a contiguous territorial unit. Indian Muslims did, however, live within such a territorial unit, and that was India. But India was not comparable to the territorial units of European nations. For Indians, and for Indian Muslims specifically, the existence of separate communities in India did not negate the possibility of building an Indian nation, but an independent India would be required to deal effectively with its own diversity if it was to successfully create the conditions whereby all of its citizens could live freely and prosperously. He said, 'The units of Indian society are not territorial as in European countries. . . the principle of European democracy cannot be applied to India without recognising the fact of communal groups' (1930, in Iqbal, 1977, p. 10). Iqbal's 1930 address, like much of Sayyid Ahmad Khan's writings on the subject of religious difference in India, ultimately worked to 'deterritorializ[e] both India and Islam', in Devji's words (2007, p. 126).

Eventually, after 1940, Muhammad Ali Jinnah began to draw more heavily upon the rhetoric of the two-nation theory in his demand for a political state for Indian Muslims. However, Jinnah's articulation of the two-nation theory transformed Iqbal's concept of identity. While Iqbal emphasised a historical religious and cultural basis for identities in India, Jinnah emphasised the translation of such identities into political and civic identities. For Jinnah, security and stability for Indian Muslims required a certain amount of territorial autonomy. Drawing on European notions prevalent at the time of what constituted a 'nation' Jinnah claimed that

> The Hindus and Muslims. . . belong to two different civilisations which are based mainly on conflicting ideas and conceptions. Their aspects on life and of life are different. It is quite clear that Hindus and Muslims derive their inspiration from

> different sources of history. . . To yoke together two such nations under a single state, one as a numerical minority and the other as a majority, must lead to growing discontent and final destruction of any fabric that may be so built up for the government of such a state. (1940, in Hay, 1988, p. 230)

Importantly, Jinnah's 'nation' differs from the 'nations' described in Hunter's early gazetteers and in Sayyid Ahmad Khan's writing. For Jinnah, the translation of religious, cultural and historical identities into the political citizenship and subjecthood of the type of nation-state India would become required homogeneity. I will return to Jinnah's emphasis on the relationship between historical religious identities and territorial organisation later in this chapter. For now, I aim only to illustrate some of the ways in which the two-nation theory has an eventful and fluid quality which is imbued with certain contemporary social and political values and debates, a feature which continues to the present day.

Problems of Scale: Difference, Representation and Electoral Politics

Sayyid Ahmad Khan's contributions to philosophy and history, and to the political debates of his times, cannot be reduced to the two-nation theory, which has its own lineage into the 21st century. Ahmad Khan was not simply concerned with religious difference, although religiosity, community and the history of Indian Muslims animated much of his thinking. He feared the electoral power of difference in the context of a state governed through representative democracy, whether that difference be linguistic, regional, class, caste or ethnic. He overwhelmingly saw the potential for difference to cause division and conflict among Indians. This fear formed the basis of his opposition to the Congress party because, he argued, the divisive power of difference was both tempered and enacted through the specificities of colonial governmentality.

Ahmad Khan was influenced in part by the philosophical tradition of English liberalism, and particularly by the work of John Stuart Mill. Ahmad Khan was deeply interested in Western modes of political organisation and governance, and doubted the potential for India to become a functional and productive democratic republic of the sort advocated by Mill.

> I reached the conclusion that the first requisite of a representative government is that the voters should possess the highest degree of homogeneity. In a form of government which depends for its functioning upon majorities, it is necessary that the people should have no differences in the matter of nationality, religion, ways of living, customs, mores, culture and historical traditions. In a country like India where homogeneity does not exist in any one of these fields, the introduction of representative government cannot produce any beneficial results; it can only result in interfering with the peace and prosperity of the land. (1888, in Hay, 1988, p. 746)

He wrote this in 1888, in response to the founding of the INC, arguing that Congress was 'based upon an ignorance of history and present-day realities; they do not take into consideration that India is inhabited by different nationalities' (in Hay, 1988, p. 747). Once again, this speaks to the contradictions inherent in the liberalist ideology underpinning colonialism in the 19th century discussed previously. While liberalism implied the dissolution of boundaries and promoted ideals of universalism, the colonial order was predicated on a rule of difference, discussed at length in the previous chapter, which proclaimed and enacted the erection and defence of borders and boundaries, both physical and social/cultural (Mehta, 1999).

But it also shows how Muslim thinkers had to reckon with the encroaching administrative and bureaucratic structures of representative government, which was being enacted in India after 1857, and which formed the basis for nationalist claims for self-rule. As Roy Bar Sadeh and Lotte Houwink ten Cate argue, Ahmad Khan was acutely aware of the 'emergence of the concept of "minority" as a key category of governance predicated on the politicisation of social difference' (2021, p. 319). In his criticism of the INC in 1887, he drew attention to the precarious position of minorities if India were to be governed through an elected governing body. He argued, 'First suppose that all the Mohammedan electors vote for a Mohammedan member and all Hindu electors vote for a Hindu member, and now count how many votes the Mohammedan members have and how many Hindu. It is certain the Hindu members will have four times as many because their population is four times as numerous... and now how can the Mohammedan guard his interests?' (1887, in McDermott et al., 2014, p. 220).

He was also careful to articulate difference not only in terms of religion but also in terms of language, region, caste and class, drawing attention to the multiple identities contained within an individual that might be mobilised for political ends. For him, these intersecting identities were all potential sources of conflict. Importantly, he spatialises these differences, arguing that scale presents an intractable problem for representative electoral government in India. In arguing against the introduction to India of a standard competitive exam for admission to the Indian civil service, he argued that not only would Muslims be unable to compete with their Hindu counterparts but that inequality existed across all salient categories of difference. He argued, 'Have Mohammedans attained to such a position as regards higher English education, which is necessary for higher appointments, as to put them on a level with Hindus or not? Most certainly not. Now, I take Mohammedans and the Hindus of our Province together, and ask whether they are able to compete with the Bengalis or not. Most certainly not' (1887, in McDermott et al., 2014, p. 219). Without either a level playing field or a homogeneous population, representative forms of governance were not viable in India.

These fundamental problems of numerical minority and spatial scale that are central to the structure of representative democracy animated debates among Indian Muslim nationalists until Partition. Sayyid Ahmad Khan's thinking presents a fundamental tension at the heart of Indian Muslim intellectual and political life: was difference (religious, regional, linguistic) inherent in Indian society, and therefore something to be embraced as a mode of understanding 'Indianness', or was difference an impediment to building a shared understanding of 'Indianness' in the first place? For Sayyid Ahmad Khan, the solution to the intractable problem of territorialising and enumerating India's multiple 'nations' was to put off representative self-rule in India until Indians were no longer at risk of governing based on differences. Faisal Devji examines how numerical minority animated Indian Muslim thought in his book, *Muslim Zion* (2013). He argues that one important reason Sayyid Ahmad Khan defended the presence of the British in India was to 'push away the problem of numbers' by appealing to 'a politics of imperial pluralism' (Devji, 2013, p. 52).

The rigid borders of the nation-state were not yet at work in the subcontinent, and so the seemingly contradictory filaments in Sayyid Ahmad Khan's political writing could more easily exist alongside one another. Such juxtapositions become all the more apparent when his ideas are placed in a wider context alongside a range of intellectual agendas which further the sense that the late 19th and early 20th century was a colonial period and laboratory within which the Indian elite refashioned colonial knowledge and experimented with Indian identity building. But for later Muslim nationalists, some of whom are introduced below, these issues of difference and scale became increasingly difficult to resolve, as electoral government took hold in India and across the world.

The difficulty of such an approach, of course, was that it required the differentiation between the historical relationships between India's various ethnic and religious groups and Indian territory. By emphasising the past of Indian Hindus (and Indian Muslims), rather than the past of Indian territory, the early nationalists struggled to devise a nationalist historiographical narrative which could overcome the differentiating and othering tendencies of such (often colonial) narratives. Geographical knowledge about India and a spatial imaginary connected to questions of borders and territory here remains *assumed*. Borders and boundaries are more imagined than real, but they are nonetheless important. For these projects to work, India had to be understood as a single spatial entity with a continuous territorial history that remained at least somewhat constant and fixed through time.

The intellectual world of colonial India during the second half of the 19th century worked to establish not only a functioning government and civil society but also to consolidate a comprehensive system of knowledge production and collection about India. In doing so, both the British colonial project, in the form of Hunter's *Imperial Gazetteer*, and Indian intellectuals and religious reformers developed a variety of historiographical accounts of India and the Indian population. With varying degrees of 'historical accuracy' (although accuracy is less interesting here than the productive legacies of such thinking), the British and Indian elites developed interwoven and competing narratives of

India's past, all of which became part of pseudo-official knowledge used to construct narratives of nationalism, identity and sovereignty in India. Some of these narratives, like Hunter's, had an explicitly and potently cartographic bent. Other narratives, like those that came from the work of Sayyid Ahmad Khan and other Indian intellectuals, imagined and represented territory, nation and community in different – more figurative – though often no less teleological ways.

In the 20th century, anticolonial nationalism among Indian Muslim politicians and writers developed along multiple lines and was characterised in large part by disagreement over how India's future government should be structured in order to protect the rights of India's minorities. Some contributions were explicitly geographical: Choudhary Rahmat Ali, credited with inventing the acronym of Pakistan and examined below, produced a small set of pamphlets depicting an imagined cartography of a future Indian subcontinent, divided into multiple nation-states based on a combination of linguistic and religious difference. Others engaged less with geographical modes of state-making and political organising but considered the contingent and ongoing process of articulating India as a political space. For example, Shaunna Rodrigues argues that the work of Maulana Abul Kalam Azad, a prominent leader of the Congress Party and critic of the idea of Pakistan, is deeply spatial: 'For Azad', she writes, 'India, as a place, was a location that was in the process of being built' (Rodrigues, 2021, p. 381).

Debates among Indian nationalists often played out in high-profile arguments about political representation: whether the Indian government should be a federal system or should have a stronger centralised core; whether religious communities should have separate electorates; how to calculate the weightage for representation, for example. Such debates were (and remain) legal and political in rhetoric, but underneath them lies implicit political geography, and so the questions of minority and majority can be usefully spatialised as part of a project to piece together a genealogy of Partition.

Sayyid Ahmad Khan's influence on 20th-century Indian Muslims, particularly in his articulation of the two-nation theory, was incredibly significant. He was not without his critics, however, and the Congress party always contained within its ranks, and often within its leadership, prominent Indian Muslim leaders and members who articulated dissenting views of India's past and future, and of what it meant to be Indian. Muslim politicians and anticolonial nationalists were working across multiple scales simultaneously in a way that reflected the unique social and historical geographies of Islam. First, they were concerned (as Sayyid Ahmad Khan was) with the well-being of India's Muslim community, and in this, they worked at local, regional and national scales. Meanwhile, Muslim intellectuals were part of wider Muslim intellectual and social networks that extended beyond the subcontinent into what was known as 'The Muslim World' (Rodrigues, 2021).

Muslim nationalists found themselves grappling with the challenge of articulating how Indian Muslims, a numerical minority who shared certain identity-based characteristics that could be distinguished from Indian non-Muslims, could be included both within the modern Indian state as citizens, and beyond India's boundaries as members of a wider Islamic community. Devji argues that this represented 'the desire of Muslim leaders to place their community in a worldwide context within which it was Hindus who were almost, but not quite, rendered into a minority' (2013, p. 69). To elaborate on Devji's point here, and to consider it in more explicitly geographical terms, minorities and majorities are not simply numerical categories, but are fundamentally spatial. That is, in order to calculate a majority, a spatial unit or area must be identified, and populations measured within it. This is not simply a minor academic point but is rather key to understanding how Indian Muslim thinkers were managing shifting population ratios across multiple territorial and administrative boundaries, and why the multiple geographical imaginaries they constructed were often competing, contradictory or lacking in consistency. Not only was the category of Muslim contingent and contested (as the historiographical literature outlined at length in the first two chapters shows), but the spatial and territorial units used to measure and mobilise Muslim majorities and minorities were themselves historically constructed and politically contingent.

The political problem of population statistics, which Ahmad Khan had identified back in the 1880s, had by the 20th century become less hypothetical, as Indians were by this time allowed to stand for certain elected roles in government. At the national scale, Muslims and other religious minorities were outnumbered by Hindus, while in certain provinces in northern India, Muslims were in the majority. Historians of Indian Muslim anticolonial national politics have shown how this problem of scale made it nearly impossible for the Muslim League to achieve electoral success across all of the Muslim-majority territorial units that comprised Indian Muslim cartographic or geographic imaginaries (e.g. Jalal, 1994). Until the mid-1940s, Punjab and Bengal were governed by powerful regional parties that did not ally themselves with the League, and which concerned themselves with maintaining strength and influence for the provinces while trying to contain power at the centre. This proved tricky for the Muslim League, which, as Ayesha Jalal and Anil Seal (1981) show, did not manage to gain electoral success among Muslim voters in Punjab until 1946, in large part because of the strength of the Unionist Party during the interwar period. The League's Muslim separatist agenda had strong backing in the United Provinces, which had a Muslim minority, but that minority was a prominent elite minority associated in large part with Aligarh, and was concerned with protecting the cultural and intellectual legacy of the Mughal court. In the late 19th century, one form this took was in the defence of the Urdu language during the Hindi-Urdu language controversy (Ayres, 2009). But, as Jalal (2000) outlines, this form of Muslim cultural identity was not universal, was regionally rooted, and not shared across all Muslim communities in British India, where regional, caste and class identities varied significantly.

The Idea and Geographical Shape of Pakistan, 1930–1947

Pakistan was, ultimately, one of the most significant geographical contradictions of Indian nationalism. It is well-known that Jinnah's earlier territorial conceptions of Pakistan envisioned a secular federalised nation-state which included the entirety of British India's Muslim-majority provinces, including the Punjab and Bengal, and that he strongly opposed the creation of any sort of religious government in either India or Pakistan. For Jinnah, the primary issue was protection and security for the Muslim minority in an independent India; and yet, in demanding Pakistan, particularly in its earlier iterations, the territory would include a substantial minority of Hindus, and in the case of the Punjab, nearly all of India's Sikh population. Additionally, in their attempts to garner popular support for Pakistan, the Muslim League drew on a populist rhetoric of religious difference and exceptionalism, making the possibility of a secular Pakistani government, which was so central to Jinnah's political vision, far more difficult to implement.

Before Jinnah introduced Pakistan into the official Muslim League agenda, the poet Muhammad Iqbal had proposed an alternative geographical solution to the issue of the Muslim minority in his presidential address at the 25th session of the Muslim League in 1930, a speech which has sparked the interest of scholars keen to make sense of the nebulous and ambiguous conceptualisation of Pakistan prior to 1947. In that speech, Iqbal presented his own vision of a future independent India, informed by the nationalist musings of Ernest Renan, the French Orientalist historian and philologist. Renan had argued, in 1882, against the idea that racial, linguistic and religious homogeneity were the foundations of a nation. Rather, a nation was 'a great aggregation of men' who, 'in sane mind and warm heart, created a moral conscience that calls itself a nation' (Renan, 2018, pp. 262–263). In Iqbal's interpretation of Renan, the formation of the nation required a 'negation' of such racial, linguistic and religious identities in favour of a 'moral consciousness', by which he meant shared memories of a 'glorious past' and a desire by all to live together in one community. However, Iqbal rejected Renan's overall conception of religion, arguing that European notions of religion did not grasp the realities of religious life in India. For Iqbal, Islam was the driving force behind the construction of the 'moral consciousness' of Indian Muslims, but that 'moral consciousness' was part of a broader Indian consciousness which was built upon religious plurality. Iqbal believed that 'the unity of an Indian nation' required 'mutual harmony and cooperation on the part of the many' without the denial or rejection of religious and cultural identities (1930, in McDermott et al., 2014, p. 489). He said, 'Communalism in its higher aspect, then, is indispensable to the formation of a harmonious whole in a country like India', distinguishing between what he identified as a 'narrow communalism' which was 'inspired by feelings of ill-will towards other communities' and a 'higher communalism' which provided the basis for 'culture' and 'consciousness' (pp. 489–490). Iqbal believed that a 'whole and harmonious India' relied upon the 'full and free development' of communities and cultures within India. He claimed, 'The units of Indian society are not territorial as in European countries. India is a continent of human groups belonging to different races, speaking different languages, and professing different religions' (p. 490). In rejecting the idea of India as a matrix of territorially constructed populations, Iqbal expressed a vision of an Indian nation that was determined first by its cultural diversity and multiple intersecting populations, rather than by its territorial unity. In other words, he was unsettling the spatial imaginary and historical narratives constructed within and through the colonial geographical archive, specifically the gazetteering projects examined in Chapter 2.

Iqbal could not, however, fully reject the role of territory in the organisation and government of the nation, and fell back upon a territorial claim within India for a distinct place reserved for Muslim political and cultural autonomy. He said, in one of the most famous lines of his address, 'I would like to see the Punjab, North-West Frontier Province, Sind and Baluchistan amalgamated into a single state. Self-government within the British Empire, or without the British Empire, the formation of a consolidated North-West Indian Muslim State appears to me to be the final destiny of the Muslims, at least of North-West India' (p. 490). He did not give this 'Muslim State' a name, and he most certainly did not apply the term 'Pakistan'. But Iqbal set out a rough idea for a 'federal India', an India composed of a collection of territorial units which would allow for the protection of minority representation and interests in government at the national level while facilitating the 'full and free development' of all communities in India at the regional and local levels (p. 489).

Iqbal knew, of course, of the importance of territorial divisions in the context of the system of representative democracy already at work in India (a prediction made by Sayyid Ahmad Khan a generation earlier), and the territorial relationship which that system initiated between the individual voter, the collective interest and the government. The battle over separate electorates had caused the biggest political division between the Congress and the League as early as 1911, a schism which had only widened as the Congress Party and Muslim League platforms were cemented and formalised over the next two decades. The electoral system designed by the government defined and structured the possibilities not only for Indians who wanted to access the government but also the possibilities for Indian nationalists to build, discursively and practically, their future state. Importantly, however, that system was not simply imposed upon an unwilling (or, for that matter, entirely acquiescent) Indian population.

That electoral system was part of a wider process of what Manu Goswami calls the 'production of colonial state space' that sought, after 1857, to administer Indian territory through a unifying and totalising understanding of India (2004, p. 9). Goswami argues that the development of Indian nationalism in the late 19th century occurred within this broader context. She argues, 'The production of colonial state space transformed the socioeconomic geography of colonial India, consolidated modalities of state power, and deepened the reach of state-generated classificatory schemes' (2004, p. 9). The colonial state apparatus and Indian nationalism were inextricably linked, and the colonial state and the single India which upper-class Hindu intellectuals called 'Bharat' existed within a mutually constitutive sphere.

Although her argument is both rigorously innovative and convincing, Goswami makes very little mention of the position or role of Muslim nationalism in that territorial development, despite the importance of Muslim contributions to Indian intellectual and political life during the second half of the 19th century. Iqbal's intervention here was not simply an oppositional challenge to more mainstream (Hindu-inflected) constructions of Indian socio-territorial unity. Rather, Iqbal was urgently highlighting the inherent contradictions of nationalising discourses which preferenced social homogeneity and territorial unity. The kind of federation Iqbal proposed has, unsurprisingly, become a galvanising point for Pakistani reformers and revisionist historians who continue to try to make sense of the contemporary Pakistani nation-state's theocratic and anti-secular government, so far from the state envisioned by its founders, Iqbal and Jinnah.

Here, however, the alternative imagined territories that framed much of the early discourse on Pakistan are useful for understanding not just the eventual creation of Pakistan, but also the geographical limitations and contradictions of the actual partitioning process in the Punjab. For Iqbal, a reorganisation of territory away from the territorial administrative system devised and implemented by the British was in the interest of all Indians, not just Muslims. Faisal Devji argues that 'the fantastic plans of Muslim leaders should perhaps be seen more as efforts to escape the status of minority, and in doing so to question that of the majority as well in a way that would eventually redefine politics itself in India' (2013, p. 69).

Just 10 years later, however, Indian nationalist discourse had changed quite dramatically. In 1935, the British government passed another Government of India Act which expanded Indian participation in government. More Indians were elected to provincial assemblies and were given more legislative powers. Some provinces were reorganised (Burma was separated from India completely), and a federal court was established. The Act had intended to establish a Federation of India, but all major Indian parties rejected the proposed federation and Independence was subsequently postponed at the start of the Second World War. The 1937 elections led to a Congress majority in many provinces, and a general failure on the part of the Muslim League, which failed to gain enough votes to form a government in any province. The Congress won nearly 45% of all seats, while the League won 6.7%. Importantly, the League did not win any seats in the Muslim-majority North-West Frontier Province and won very few seats in the Muslim-majority United Provinces (Jalal, 1994).

Jinnah himself was notoriously wary of religious nationalism until the final decade of the Raj. A staunch secularist throughout the 1910s and 1920s, he believed that religiously infused nationalism was, in fact, a cause of the failure of democracy to protect all Indians equally. He tried, albeit ultimately unsuccessfully, to distinguish between religious identities and political identities. He believed that the categories of 'Hindu' and 'Muslim' in the political arena were distinct from the categories of 'Hindu' and 'Muslim' in the arena of personal and communal faith. For Jinnah, the specificities of the Hindu and Muslim practicing their faith at home with their families and in their places of worship should not be subject to political intervention, nor should these personal religious identities be used to mobilise and unify Indians within a shared political consciousness. In such an agenda, Jinnah saw only division and danger, specifically for those who would be excluded from such a political identity, and, more importantly, who would be in a numerical minority and thus subject not only to the political whims of a majority but also to politicised religious sentiments bolstered by the power of the state. Political identities became increasingly relevant in the final decades of the Raj, and electoral politics became the established mode of accessing formal political power in the Indian government. But as the Congress Party moved towards secularism informed more and more by the rhetoric and philosophy of Hinduism, Jinnah's secular distinction between private 'Muslims' and political 'Muslims' became untenable. The conflation of political and religious identities in Indian politics made Jinnah's secularism too academic and intellectual to mobilise followers to his cause.

The question of nationalism and its relationship to communalism has been taken up, particularly by members of the Subaltern Studies group. In the Introduction, and in Chapter 1, I drew on the work of a number of authors to illustrate the ways in which identity categories, particularly religion, were discursively articulated, enumerated, classified and disseminated through what Bernard Cohn (1996) has called 'modalities'. Religion in particular formed the basis of much of the colonial state's understanding of its Indian subjects, especially in the context of violence and disorder. In *The Construction of Communalism in Colonial North India*, Gyanendra Pandey (2012) argues the British colonial government in India, through its historiographical construction of the Indian past, developed a cohesive narrative of Indian communal violence which could explain both historical instances of civil unrest, and then-current instances of rioting, violence and disorder. For Pandey, the rigidity of the religious categories, and the insistence by the colonial state that Indian violence was 'caused' by inherent and long-standing friction between the members of different religions, created a situation whereby the concept of 'communalism' in Indian society and Indian history was folded into nationalist and imperialist discourses without ever facing a serious intellectual challenge. Communalism, therefore, like nationalism, came to be a self-evident concept in India, one with specific and observable causes and effects. Pandey argues that in nationalist negotiations in India, nationalism and communalism worked in opposition to one another, where nationalists sought to 'overcome' the lesser, inferior pull of communalism in favour of an all-Indian secular politics.

Partha Chatterjee explains this opposition in terms of the colonial state's assumption that '"communities" rather than "nation" were what characterised [Indian] society', and that 'those communities had to be singular and substantive entities in themselves, with determinate and impermeable boundaries, so insular in their differences with one another as to be incapable of being merged into larger, more modern political identities' (2007, p. 224). Meanwhile, he argues, 'Nationalists, of course, rejected this' (2007, p. 224). Here, I follow Pandey and Chatterjee in arguing that nationalism and communalism were in tension in India before 1947, but that the idea does not fit so neatly when applied specifically to Jinnah, Iqbal and the geographical imaginary of Pakistan after 1930. This is partly to do, of course, with the postcolonial nature of this project. Pakistan and India have existed as independent nation-states for over 75 years, and, as Chatterjee (2007), Pandey (2001), Chakrabarty (2000) and Guha (2002) (among others) remind us, writing the histories of those states is always 'conditioned and limited' (Chatterjee, 2007, p. 77). But this literature, and indeed much of the literature produced by members of the Subaltern Studies group, has, despite their myriad ground-breaking and vital intellectual contributions, been unable to fit a narrative of Pakistan, at least a narrative that is as fully developed as that of India, into their theories of nationalism in any sustained way.

The question of communalism and nationalism is central to histories of Partition, especially those older, more traditional narratives which position Jinnah and the Muslim League as being in favour of Pakistan, and Nehru and the Congress as being against Pakistan (a position that revisionist work has challenged). Yet, if nationalism and communalism worked in opposition to each other, how might we explain the uneven and difficult negotiation between the two forces within the official discourses of the Muslim League leadership, and the Muslim intellectual elite after 1930, especially with regard to the spatial imaginaries of Pakistan? Was Pakistan a nationalist ideal, or a communal ideal? Was the League's brand of nationalism fundamentally different from that of the Congress? Or were they variations of the same overall elite nationalist framework? Chatterjee (2007) argues that the postcolonial Indian state, despite the general nationalist rejection of colonial categories, was unable to free itself from that construct. He notes, 'An underlying current of thinking about the sociological bases of Indian politics continues to turn along channels excavated by colonial discourse' (Chatterjee, 2007, p. 224). This theme runs through much of postcolonial studies and informs much of the argument here. But did this process work the same way in both India and Pakistan? Certainly, Pakistan and India have both been shaped by continued negotiations between religion, particularly fundamentalist religion, and governance. But is Pakistan subject to different rules because the nationalism of the Muslim League was fundamentally different from that of the Congress? Pandey and Chatterjee are not necessarily interested in teasing out the ways in which Indian history is prioritised by studies of colonial and postcolonial nationalism in the subcontinent.

Choudhary Rahmat Ali and the 'Continent of Dinia'

Importantly, from a cartographic point of view, before 1940, Jinnah and the Muslim League never officially invoked an image of an independent India that looked particularly different from the India imagined by the other nationalist parties. The Muslim League was concerned primarily with redefining the *relationship* between the individual, the place in which they lived and the state, both through a re-fashioning of the electoral system and a reimagining of India's *internal* boundaries. It was not until 1940 that this position shifted, and moved towards a serious discussion of partitioning India, enacting an *external* boundary and creating two new nation-states. This process was partly characterised by a renegotiation between the related but distinctive discourses of communalism and nationalism by the League elite (discussed above), and most especially Jinnah. In March 1940, Jinnah addressed the Muslim League in Lahore, for the first time calling for an explicitly independent territorial homeland, legally and territorially separate from India, for Indian Muslims.

'Pakistan' was first formally articulated in 1933 by Choudhary Rahmat Ali, a Punjabi Muslim student of law at Cambridge University, in a pamphlet titled 'Now or Never: Are We to Live or Perish Forever?', written for the Round Table Conference that took place in London. Rahmat Ali called himself the 'Founder of the Pakistan Movement', and he developed his nationalist philosophy during the 1930s in a series of pamphlets and articles. In 1934, Jinnah spoke at Cambridge and was approached by Rahmat Ali, who introduced Jinnah to his idea of Pakistan. Jinnah famously rejected it, but Rahmat Ali continued to write about and imagine a territorial homeland for Muslims in India. There are relatively few cartographic representations of Pakistan before 1947 and many of those that exist were produced by Rahmat Ali. This is unsurprising, of course, given Jalal's argument that a Pakistan contained within a bounded and contiguous territory was not Jinnah's primary goal at all, but was in fact a political bargaining chip. Rahmat Ali's term 'Pakistan' is, famously, an acronym for those territorial regions within India which were heavily populated by Muslims, and which were central to Indian Muslim history. As he wrote: 'PAKSTAN, by which we mean the five Northern units of India viz: **P**unjab, North-West Frontier Province (**A**fghan Province), **K**ashmir, **S**ind, and Baluchis**tan**', (1933, no pagination). It is notable that Kashmir is included in Rahmat Ali's PAKSTAN, while Bengal is not, and is instead represented as a separate independent state, Bang-i-Islam, a territorial unit created not just on the basis of Muslim majority, but also on the basis of a shared language, Bengali. While 'Pakstan' was constructed out of regions where Muslims represented majority populations but spoke a diversity of languages, Bang-i-Islam also demonstrated Rahmat Ali's identification of shared language as a key feature of the nation. But the inconsistency (shared language in Bengal, diverse languages in the north and west) reflects the wider challenges faced by the Muslim League and Indian Muslim nationalists outside of the League when they were called upon to articulate how exactly India's Muslim communities were united in cultural, social, political or linguistic terms, and how these characteristics were distinct from non-Muslim Indians. It also helps explain why the idea of Pakistan did not gain traction among Muslims in those very provinces, especially in Punjab. Language was also significant in those provinces that made up Rahmat Ali's 'Pakstan', as Farina Mir (2010) shows in her work on Punjabi, and its role in fashioning a distinctive Punjabi cultural and social world that cut across communal distinctions.

Alyssa Ayres shows how this works in post-Independence Pakistan, using language as a lens to argue that Pakistan was 'far more than just an acronym presented by Rahmat Ali. It was a lexeme for population unmixing, a cleansing, that the land itself would have to undergo to become pure and pāk' (2009, p. 27). She draws on how Rahmat Ali imbued Pakistan with a double meaning through his use of the word 'pāk', which 'in Islamicate languages of Northern India via Persian. . .denotes purity, virtue, even holiness' (2009, p. 26). In practical terms, such an 'unmixing', if it were to go beyond abstraction or symbolism, would have to be reflected by population statistics, which required what the mainstream League leadership always tried to deny, even after 14 August 1947: that Pakistan would both instigate and require a large-scale transfer of populations. This account was being developed at the same time as Anglo-European perspectives on geographical boundaries that were arguing for the opposite: that population transfer should be avoided in boundary-making exercises, except in the most extreme circumstances (Jones, 1945).

Inspired by Iqbal's presidential address, Rahmat Ali adapted Iqbal's notion of a large Muslim-administered federal state and argued that Pakistan was, in fact, a distinct (and notably contiguous) territorial unit based on Indian Muslims being a distinct and separate social and cultural nation. In this regard, the specificity and enthusiasm with which he envisioned the subcontinent was relatively rare and ambiguous (Chester, 2019). But Rahmat Ali's pamphlets were not just maps of independent states in the subcontinent. Like all anticolonial nationalists, he narrated a need for such new boundaries using a discourse of historicality. He explicitly linked the League's political rhetoric about Muslims' distinct political interests with Iqbal's rhetoric about Muslims' distinct spiritual and historical development to fashion a claim for a territorial nation-state which he projected back in time and imagined into the future, reflecting some of Anderson's (2006) observations on discursive constructions of the nation-state.

Anderson, in *Imagined Communities*, emphasised the tendency for nationalists to articulate the nation 'genealogically as the expression of a historical tradition of serial continuity' (2006, p. 195). Anderson argued, in an updated edition of this book, that anti-imperial nationalism in the colonies was necessarily shaped by 'imaginings of the colonial state', despite the general fact that colonial states were opposed to nationalism (in the colonies. In the metropole, colonialism worked in many ways to construct European nationalism in the 19th century)

(2006, p. 163). Rahmat Ali's particular notion of 'Dinia' relied both on a constructed 'timelessness' of the diversity and heterogeneity (in the form of multiple distinct 'communities') of the Indian population, and on the colonial categories which articulated, enumerated and spatially located those communities.

For Rahmat Ali (and for many other separatist nationalists), the cultural and social distinctions which existed between Hindus and Muslims should be matched by a separation in territorial affiliation. He believed that territorial boundaries should match the social and political boundaries already at work in Indian society. This was not, of course, so different from the goals of the British government during the second half of the 19th century. The purpose of the gazetteer atlas maps, discussed in Chapter 2, was to visually and cartographically represent territorial divisions that corresponded with social and religious divisions in Indian society, to construct a socio-territorial grid or matrix. In his first pamphlet, Rahmat Ali wrote, 'India, constituted as it is at the present moment, is not the name of one single country; nor the home of one single nation. It is, in fact, the designation of a state created by the British for the first time in history' (1933, no pagination).

As we have seen, Hunter's *Imperial Gazetteer* similarly understood India as being populated by many different nations and races, all of whom had been brought together under the unifying administration of the British. Rahmat Ali continued, 'In the five Northern Provinces of India, out of a total population of about forty millions, we, the Muslims, constitute about thirty millions. Our religion and culture, our history and tradition, our social code and economic system, our laws of inheritance, succession and marriage are fundamentally different from those of most peoples living in the rest of India' (1933, no pagination). He believed, like Jinnah, and like Sayyid Ahmad Khan before him, that the position of an electoral minority was dangerous for Indian Muslims. Unlike Ahmad Khan, he believed in independence from the British, and unlike Jinnah, he did not believe that such a minority could be compensated for by political safeguards.

Rahmat Ali appealed to the scale and concentration of the Muslim population to argue that the social separations between Hindus and Muslims *could already be seen on the map*. He argued: 'The total area of our [Muslim] five units, comprising Pakistan, is four times that of Italy, three times that of Germany and twice that of France; and their population seven times that of the Commonwealth of Australia, four times that of the dominion of Canada, twice that of Spain, and equal to France and Italy considered individually' (1933, no pagination).

Until recently, historians have tended to overlook the fact that Rahmat Ali expanded on his cartographic vision for an independent subcontinent, drawing on his interpretation of India's multinational character and diversity of populations to argue that separate 'homelands' for all potentially marginalised populations in India were the most effective solution to communal, caste and racial tensions. Alyssa Ayres (2009) focuses on his use of shared language as a defining characteristic of the nation. Meanwhile, Tahir Kamran (2017) has examined historically understudied papers produced by Rahmat Ali to show how he was influenced by political and social conditions in Punjab, and how they evolved during the 1930s and 1940s. And Sumathi Ramaswamy has briefly but effectively invoked his novel depiction of 'Akhootistan', or 'Land of the Achuts, or Untouchables' on his map of Dinia (Ramaswamy, 2012, p. 206).

Rahmat Ali renamed the subcontinent Dinia, saying

> whereas the word "India" defines the lands as the exclusive domain of Caste Hindooism and Caste Hindoos and consequently denies the existence and share therein of Dravidianism and Dravidians, of Akhootism and Akhoots, of Buddhism and Buddhs, of Islam and Muslims, of Sikhism and Sikhs, of Christianity and Christians, of Zoroastrianism and Parsis, and misrepresents all peoples as Caste Hindoos in the lands of Caste Hindooism; the word "Dinia" defines these lands as the joint domain of all the religions and their followers found therein, and consequently acknowledges the existence and share therein of them all, and describes them as the peoples of the lands of religions without reference to any particular religion or fraternity. (1945, p. 8)

He proposed territorial homelands for each group within 'Dinia', arguing with geographical logic that the subcontinent was, in fact, a 'continent' and could *not* be called a 'country'. As such, there was no need for India to be granted Independence as a single political unit, given that it was made up of so many distinct nations. In arguing this point, he appealed to a simple interpretation of geography, which he outlined in a 1945 pamphlet titled 'India: The Continent of Dinia, or the Country of DOOM'. A country, he said, was 'just a fair sized politically demarcated area of the land that possesses some individual characteristics' while a continent 'is a huge, continuous mass of land that is bordered by mountain chains or high seas, or partly by one and partly by the other'. He said that 'if we open our atlases and in the light of these broad, basic definitions, look at the map of India', one would see that, as many Indian intellectuals pointed out in various contexts, India is an area that is 'at least equal to the whole Continent of Europe, excluding Russia' (1945, p. 7). India was bounded, like Hunter had noted in the *Imperial Gazetteer*, 'in the north-east, north, and north-west, it is shielded by the highest mountain ranges in the world; and, in the south-east, south, and south-west, its shores are washed by a vast ocean and high seas' (1945, p. 7).

In fact, Rahmat Ali may very well have devised Dinia not just from Iqbal's earlier conceptualisation, but also from older medieval maps produced by Mughal scholars, which often represented India in terms of multiple states and principalities. There is not scope within this book to examine in detail pre-colonial forms of mapping, and medieval modes of territorial thinking in India. However, it is important to note that cartographic and geographical thinking existed in India before the arrival of the British, and early colonial depictions of India were influenced by some of these sources. During the later colonial period, many of these sources were forgotten or ignored by colonial writers like Hunter, who were constructing a totalising narrative of Indian unity under British rule. Later nationalist thinkers engaged in a number of projects to recover and mobilise older Indian forms of geographical and political thought for the purpose of reimagining India and Pakistan as independent political and territorial units. This is, of course, an important part of Rahmat Ali's Dinia project.

Rahmat Ali believed that the diversity in Indian geography and population, which the British believed was their responsibility to understand and govern, was evidence that India should be divided into multiple independent territorial-political units along religious lines. He proposed a new territorial connection that rejected the idea of 'India' to unify these many 'nations': that of 'Dinia', a 'continent' not a 'nation'. Rahmat Ali depicted this territorial reordering using maps, showing the boundaries within 'Dinia', what he called 'a continent of sovereign nations, living in separate homelands, working out their individual national destinies in their own ways, and making their national contributions to the solution of the common problems of Dinia, of Asia and the world' (1945, p. 11). The most famous is a map titled 'Continent of Dinia and its Dependencies' (Figure 3.1), which attempts to show what he called the 'Myth of Indianism' (no pagination). The map depicts Pakistan, of course, which occupies the north-west corner of the subcontinent, and Bangistan, which occupies the north-eastern corner, in the province of Bengal. Interestingly, not all of the green regions had a Muslim majority. Osmanistan, for example, was a landlocked block in the centre of the subcontinent, representing Rahmat Ali's vision for the post-Independence princely state of Hyderabad, a Hindu-majority state ruled by a Muslim prince, the Nizam Osman Ali Khan. Osmanistan, according to Rahmat Ali, should have been an independent state with the same claim to legitimacy as Pakistan, based on historical territorial claims to sovereignty. Rahmat Ali was interested in constructing what Martin W. Lewis has called in an online comment piece a 'new kind of political-cultural space' where India's minorities could be included and invested in a continental community based on a shift in the geographical conception of India (Lewis, 2010, no pagination). Lewis (2010) argues that Dinia is interesting primarily because activists in Pakistan have used the concept to argue that Muslims were wronged in 1947, and that the Pakistan which exists today does not match Rahmat Ali's vision. I do not use Dinia in this way here. Rather I aim to show the multiplicities of Muslim nationalist discourse in the final decades of the Raj and the ways in which the Pakistan idea inspired various imagined geographies before its eventual creation.

Like all nationalists, Rahmat Ali drew upon a creative combination of local and colonial knowledge to construct a historiography of Muslims in India which linked their political interests with his proposed 'Pakistan', thereby constructing a Muslim territorial claim. For example, he drew on the sensitive political topic

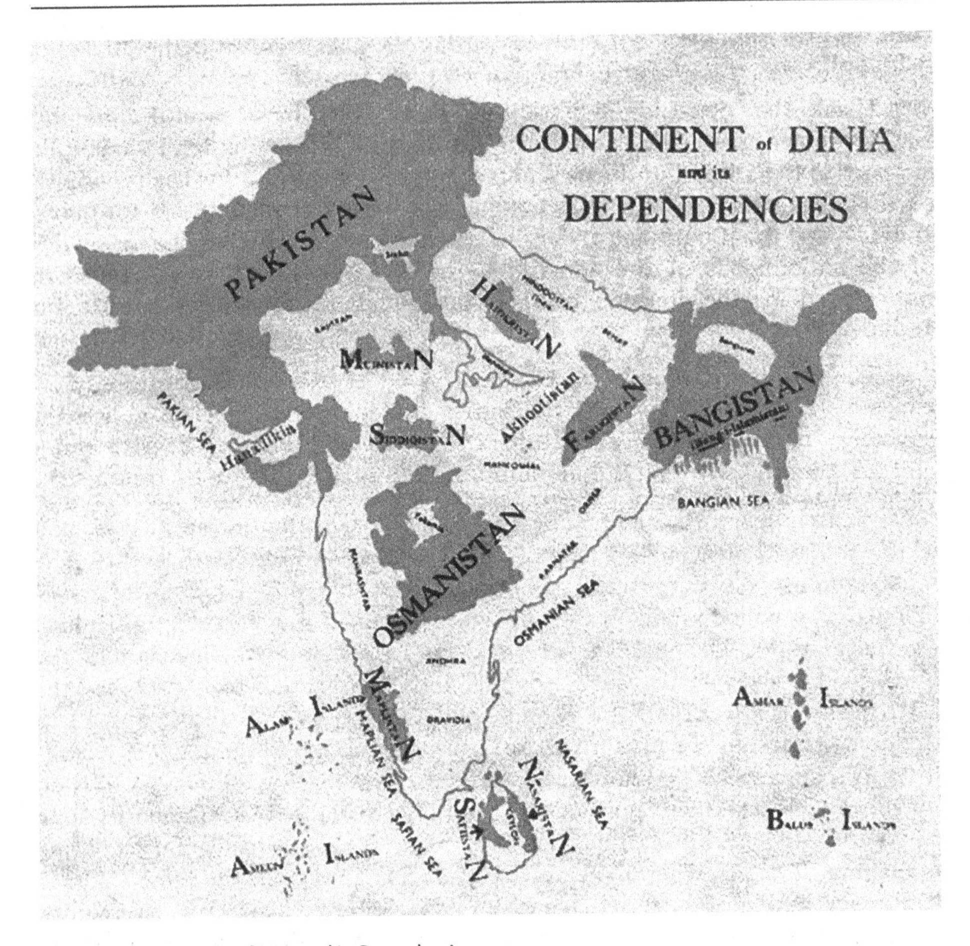

Figure 3.1 Continent of Dinia and its Dependencies.
Source: Choudhary Rahmat Ali, 'INDIA The Continent of DINIA or The Country of DOOM?,' 1945. Source: National Library of Scotland.

of the Urdu language, believed by many concerned Muslims to have been politically marginalised, to illustrate the cultural and social threat that a united India posed to Muslims in India: 'The very name of our national language – Urdu, even now the *lingua franca* of that great sub-continent, has been wiped out of the list of Indian languages. We have just to open the latest census report to verify it' (1933, no pagination). Like the reformers and early nationalists of the 19th century, Rahmat Ali and other nationalists relied on official colonial records to illustrate their essential and historical differences, while at the same time working to reject other colonially constructed forms of knowledge.

The contradiction which characterises all nationalist re-imaginings of colonial knowledge exists here, of course; Rahmat Ali relied on British geographical knowledge, collected over almost two centuries via surveys and maps, in order to reject the British interpretation of that knowledge. He mobilised those knowledge categories which had been devised and utilised by the British colonial government, especially linguistic and religious categories. Unlike other nationalists in India, however, Rahmat Ali believed that the unity which the British had imposed on Indian geography and Indian history was, in fact, a colonial construction, and that, if India were to move towards a political system which matched its geography, it should become many independent states, based on those same religious and linguistic communities (see Figure 3.1, map of 'Continent of Dinia'). In this way, Rahmat Ali imagined a different specific political geography out of broader colonial forms of knowledge about India.

Rahmat Ali and Iqbal appealed to the same geographical arguments over security and stability that underpinned much of the colonial administrative discourse regarding the management of Indian territory. Rahmat Ali spoke to the potential for Pakistan to act as a territorial 'buffer state' between India and the rest of Asia, reflecting British colonial ideas about the management of territory for the purpose of security. Thomas Holdich, introduced in the previous chapter, wrote extensively on the role of buffer states in establishing regional security. According to Rahmat Ali, 'this Muslim Federation of North-West India would provide the bulwark of a buffer state against the invasion of India either of ideas or of arms from any quarter' (Ali, 1933). Iqbal, in his address to the Muslim League in Allahabad, noted the importance of the region as a border region, which would require the Indian government to take over from the British the task of border policing and surveillance after Independence. He proposed 'a strong Indian Frontier Army', to be 'composed of units recruited from all provinces and officered by efficient and experienced military men taken from all communities' (1930, in Iqbal, 1977, p. 19). Of course, neither took into account the possibility of adjacent nation-states becoming a security threat to each other, which of course is what transpired.

Jinnah and the Lahore Declaration

As mentioned above, 'Pakistan' remained a marginal idea within Indian Muslim nationalist discourse until 1940, and 'Dinia' never made a meaningful appearance in mainstream Indian nationalist discourse. But, in 1940, in the 'Lahore Declaration' (also called the 'Pakistan Declaration', although there is no mention of Pakistan within it), Jinnah explicitly linked the future territorio-political organisation of Indian Muslims with the more radical Muslim nationalist conception of separatism, and the establishment of a political, spiritual and territorially bounded homeland for Muslims in India. He said, 'Mussalmans are a nation according to any definition of a nation, and they must have their homelands, their territory, and their state' (in McDermott et al., 2014, p. 503). Of course, Jinnah's new rhetoric reflects that of Sayyid Ahmad Khan and other Muslim nationalists in its conceptualisation of Muslims and Hindus as deriving their identities from parallel but distinct sets of social rules, values and history. Whatever his reasons for moving towards a call for an independent geographical homeland for Muslims in India, and again I recall here Jalal's (1994) argument that Pakistan was Jinnah's 'bargaining chip', Jinnah struggled with both the territorial limitations imposed by the Congress platform of Indian unity and the standards in boundary-making and ethnographic mapmaking of the time. In order to galvanise Muslim support for the idea of Pakistan in the first place, Jinnah and the Muslim League had to frame the concept as a 'homeland for Indian Muslims', evoking an affective historico-religious form of ethnic nationalism which was not necessarily a viable representation of the geopolitical possibilities provided by standard practices in boundary-making. Imagined homelands and territorial nation-states rarely reflect one another in practice, but they continue to exist alongside each other, often causing friction between groups, and within the state and sections of the population.

In 1946, while Lord Wavell was still Viceroy of India, the final attempt to resolve the protracted conflict between the Congress and the League over the British transfer of power to India failed. The Cabinet Mission Plan, ultimately rejected by Nehru and the Congress, would have created a federation of states with a relatively less powerful central government than Nehru was willing to concede. In 1947, Lord Mountbatten was sent to India with the directive to transfer power by the summer of 1948. Soon after his arrival, Mountbatten declared that power would be transferred a year early, and set about devising the terms and conditions under which independence would be granted. By 3 June, it was agreed that a partition would be implemented as the solution to the intractable negotiations between the League and the Congress.

Despite the fact that 'Pakistan' did not necessarily imply a territorial partition until after 1940, and even then, the League leadership was willing to compromise on its final form when they accepted the Cabinet Mission Plan in 1946, the terms 'Pakistan' and 'Partition' are now deeply intertwined in imperial and Indian nationalist historiographies. Even now, the creation of the state of Pakistan and the process by which the border between India and Pakistan was devised are closely connected. This has the effect, of course, of undermining most official Pakistani narratives of Independence, but also of cementing not only onto the world map but also into historical memory, one particular geographical representation of the subcontinent. Before 1947, Pakistan and India were at once a British colony on the verge of independence and a set of imagined geographies, wherein historical and cultural debates about Indian identities and visions of the future were let loose to do a kind of affective-intellectual battle. After 1947, political discourse was focused on consolidating official narratives of the creation of the new nation-states, while on the ground, individuals and communities went about making sense of their new political and cultural identities.

Conclusion

In the final decades of the Raj, Muslim nationalist discourses drew on a number of influences, both British and Indian. I have tried to show here how the territoriality and historicality of these discourses were informed by older colonial modes of understanding and ordering space, as well as by the ways in which earlier Indian reformers and intellectuals had articulated their own conceptualisations of India. I have focused on the idea of Pakistan here primarily because Pakistan presented a series of challenges to the notion of Indian unity, and I argue that by understanding the idea of 'Pakistan' as a form of reterritorialising Indian nationalism, we can open up some of the ways in which partitions and boundary-making were a necessarily incomplete and imperfect fix to the geopolitical process of decolonisation in India.

'Pakistan' was a nebulous and ambiguous concept, one which was refracted through many competing intellectual 'lenses': that of the idealistic intellectual student, for example, and the religious philosopher, the pragmatic politician, and, eventually, the colonial official. My arguments here are not an attempt to challenge historical narratives of the causes and effects of Partition, or of the historical explanations for why Pakistan took the eventual shape it did. Rather, I seek to open up a debate about the spatialising tendencies of Indian nationalism, and the ways in which Indians imagined the future independent India. The idea of Pakistan was very much rooted in the geographical and cartographic authority of imperial knowledge and was refracted through older intellectual engagements with that knowledge. The *Imperial Gazetteer of India*, discussed at length in Chapter 2, had, by 1930, become the geographical authority in India, while the Survey of India had become the primary producer of maps and surveys. Yet Pakistan was also a challenge to dominant nationalist modes of framing India as a single territorial unit, and the Indian people as a unified group of people with a set of shared interests that could and should be protected by the Congress secular democratic model. Pakistan was a challenge to the notion that electoral politics, at least without safeguards for minorities, was the best kind of politics for India, and that a political and social ideal of unity despite diversity was in the interest of all Indians.

The territorial reality of Pakistan (and of India) only came into being after Partition in 1947. The partition itself was a relatively mundane geopolitical process, carried out by the middling elite – judges, lawyers, and career politicians – on behalf of the elite leadership and the colonial government. As such, it was one of the first large-scale processes by which the territory of the colonial world was reorganised cartographically and institutionally. Yet the process could not entirely reject the modes of geographical organisation and cartographic representation that had built the colonial world order, and as such, marked an asymmetrical transition to territorial and political independence. Scholars from the Subaltern Studies school have pointed to the failures of European modes of political and social organisation (democracy and capitalism, for example) to effectively produce their philosophical ideals (political and economic freedom) in the Indian context. I do not argue here that there was necessarily an indigenous 'Indian' mode of territoriality that would have provided a more effective solution than a partition to the protracted communal tensions, particularly in the Punjab and Bengal. Rather, I hope to point to the discursive and epistemological limits of cartography and boundary-making in the context of Partition, bound up as they were in the ordering of the colonial world and the imperial worldview. The end of empire was in many ways an unravelling of the imperial geographical and cartographic representation of the world, and the redrawing of a new postcolonial map. Territorial partitions were part of this process.

In the next two chapters, I turn to the Punjab Boundary Commission hearings, where many of these imagined geographies of the future independent subcontinent were introduced to the high stakes of the courtroom, and the era of British colonialism drew to a close.

References

Ali, C.R. (1933). *Now or Never, Are We to Live or Perish Forever?* Cambridge: The Pakistan National Movement.

Ali, C.R. (1945). *INDIA The Continent of DINIA or The Country of DOOM?* Cambridge: The Dinia Continental Movement.

Anderson, B. (2006). *Imagined Communities: Reflections on the Origin and Spread of Nationalism*. New York; London: Verso.

Ansari, S. (1992). *Sufi Saints and State Power: The Pirs of Sind, 1843–1947*. Cambridge: Cambridge University Press.

Asif, M.A. (2020). *The Loss of Hindustan: The Invention of India*. Cambridge, Massachusetts: Harvard University Press.

Aydin, C. (2017). *The Idea of the Muslim World: A Global Intellectual History*. Cambridge, Massachusetts: Harvard University Press.

Ayres, A. (2009). *Speaking Like a State: Language and Nationalism in Pakistan*. Cambridge: Cambridge University Press.

Bar Sadeh, R. and Houwink ten Cate, L. (2021). Toward a Global Intellectual History of 'Minority'. *Comparative Studies of South Asia, Africa and the Middle East* 41 (3): 319–324. https://doi.org/10.1215/1089201X-9407819.

Basu, S. (2010). The Dialectics of Resistance: Colonial Geography, Bengali Literati and the Racial Mapping of Indian Identity. *Modern Asian Studies* 44 (1): 53–79.

Bayly, C.A. (2011). *Recovering Liberties: Indian Thought in the Age of Liberalism and Empire*. Cambridge: Cambridge University Press.

Bose, S. (2007). The Spirit and Form of an Ethical Polity: A Meditation on Aurobindo's Thought. *Modern Intellectual History* 4 (1): 129–144. https://doi.org/10.1017/S1479244306001089.

Chakrabarty, D. (2000). *Provincializing Europe: Postcolonial Thought and Historical Difference*. Princeton, New Jersey: Princeton University Press.

Chatterjee, P. (2007). *The Partha Chatterjee Omnibus: Nationalist Thought and the Colonial World, The Nation and Its Fragments, A Possible India*. New Delhi; New York: Oxford University Press.

Chester, L. (2019). Image and Imagination in the Creation of Pakistan. In *Mapping Migration, Identity, and Space*, edited by T. Linhard and T.H. Parsons, 137–158. Cham: Springer International Publishing.

Cohn, B. (1996). *Colonialism and Its Forms of Knowledge: The British in India*. Princeton, New Jersey: Princeton University Press.

Cox, J. (2002). *Imperial Fault Lines: Christianity and Colonial Power in India, 1818–1940*. Stanford, California: Stanford University Press.

Devji, F. (2007). A Shadow Nation: The Making of Muslim India. In *Beyond Sovereignty: Britain, Empire and Transnationalism, c. 1880–1950*, edited by K. Grant, P. Levine, and F. Trentmann, 126–145. London: Palgrave Macmillan UK. https://doi.org/10.1057/9780230626522_7.

Devji, F. (2013). *Muslim Zion: Pakistan as a Political Idea*. Cambridge, Massachusetts: Harvard University Press.

Devji, F. (2014). India in the Muslim Imagination: Cartography and Landscape in 19th Century Urdu Literature. *South Asia Multidisciplinary Academic Journal*, 10 (December). https://doi.org/10.4000/samaj.3751.

Gilmartin, D. (1979). Religious Leadership and the Pakistan Movement in the Punjab. *Modern Asian Studies* 13 (3): 485–517.

Goswami, M. (2004). *Producing India: From Colonial Economy to National Space*. Chicago: University of Chicago Press.

Gould, W. (2020). Social and Religious Reform in 19th-Century India. *Oxford Research Encyclopaedia of Asian History*. https://doi.org/10.1093/acrefore/9780190277727.013.382.

Guha, R. (2002). *History at the Limit of World-History*. New York: Columbia University Press.

Guha, R. (2011). *Dominance Without Hegemony: History and Power in Colonial India*. Cambridge, Massachusetts: Harvard University Press.

Hasan, M. (1993). Resistance and Acquiescence in North India: Muslim Responses to the West. *Rivista Degli Studi Orientali* 67 (1/2): 83–105.

Hay, S. (1988). *Sources of Indian Tradition, Vol. 2: Modern India and Pakistan*, 2nd ed. New York: Columbia University Press.

Jalal, A. (1994). *The Sole Spokesman: Jinnah, the Muslim League and the Demand for Pakistan*. Cambridge: Cambridge University Press.

Jalal, A. (2000). *Self and Sovereignty: Individual and Community in South Asian Islam Since 1850*. London: Routledge.

Jalal, A and Seal, A. (1981). Alternative to Partition: Muslim Politics Between the Wars. *Modern Asian Studies* 15 (3): 415–454. https://doi.org/10.1017/S0026749X00008659.

Jones, K. (1989). *Arya Dharm: Hindu Consciousness in 19th-Century Punjab*. Delhi: Manohar.

Jones, K. (1992). *Religious Controversy in British India: Dialogues in South Asian Languages*. Albany: SUNY Press.

Jones, K. (2006). *Socio-Religious Reform Movements in British India*. New York: Cambridge University Press.

Jones, S. (1945). *Boundary-Making: A Handbook for Statesmen, Treaty Editors and Boundary Commissioners*. Washington: Carnegie Endowment for International Peace.

Kamran, T. (2017). Choudhary Rahmat Ali and His Political Imagination: Pak Plan and the Continent of Dinia, in *Muslims Against the Muslim League: Critiques of the Idea of Pakistan*, edited by A.U. Qasmi and M.E. Robb Cambridge: Cambridge University Press.

Lewis, M.W. (2010). Dreams of Dinia and a Greater Pakistan. *GeoCurrents*. http://www.geocurrents.info/geopolitics/dreams-of-dinia-and-a-greater-pakistan.

Majeed, J. (2007). *Autobiography, Travel, and Postnational Identity: Gandhi, Nehru and Iqbal*. Basingstoke: Palgrave Macmillan.

Majeed, J. (2009). *Muhammad Iqbal*. New Delhi: Routledge, Taylor & Francis Group.

McDermott, R., Gordon, L.A., Embree, A.T., et al. eds. (2014). *Sources of Indian Tradition: Modern India, Pakistan, and Bangladesh*. New York: Columbia University Press.

Mehta, U.S. (1999). *Liberalism and Empire: A Study in Nineteenth-Century British Liberal Thought*. Chicago: University Of Chicago Press.

Metcalf, B. (1978). The Madrasa at Deoband: A Model for Religious Education in Modern India. *Modern Asian Studies* 12 (1): 111–134.

Metcalf, B.D. (1982). *Islamic Revival in British India*. Princeton, New Jersey: Princeton University Press.

Mir, F. (2010). *The Social Space of Language: Vernacular Culture in British Colonial Punjab*. Berkeley: University of California Press.

Pandey, G. (2001). *Remembering Partition: Violence, Nationalism and History in India*. Cambridge: Cambridge University Press.

Pandey, G. (2012). *The Construction of Communalism in Colonial North India*, 3rd ed. Delhi: OUP India.

Qasmi, A.U. (2014). *The Ahmadis and the Politics of Religious Exclusion in Pakistan*. Cambridge Core. Anthem Press. May 2014.

Ramaswamy, S. (2002). Visualising India's Geo-Body Globes, Maps, Bodyscapes. *Contributions to Indian Sociology* 36 (1–2): 151–189. https://doi.org/10.1177/006996670203600106.

Ramaswamy, S. (2009). *The Goddess and the Nation: Mapping Mother India*. Durham, North Carolina: Duke University Press.

Ramaswamy, S. (2012). The Work of Goddesses in the Age of Mass Reproduction. In *Transcultural Turbulences*, edited by C. Brosius and R. Wenzlhuemer, 191–220. Heidelberg: Springer.

Ramaswamy, S. (2017). *Terrestrial Lessons: The Conquest of the World as a Globe*. Chicago: University of Chicago Press.

Raza, A. (2020). *Revolutionary Pasts: Communist Internationalism in Colonial India*. Cambridge: Cambridge University Press.

Renan, E. (2018). What Is a Nation?. In *What Is a Nation? and Other Political Writings*, M.F.N. Giglioli, 247–263. New York: Columbia University Press.

Rodrigues, S. (2021). The Place of Political Membership: Abul Kalam Azad's Critique of Borders and Nations. *Comparative Studies of South Asia, Africa and the Middle East* 41 (3): 378–388. https://doi.org/10.1215/1089201X-9407923.

Iqbal, M. (1977). 1930 Presidential Address to the 25th Session of the All-India Muslim League, Allahabad, 29 December 1930. In *Speeches, Writings, and Statements of Iqbal*, edited by L.A. Sherwani, 2nd ed., 3–26. Lahore: Iqbal Academy.

Chapter Four
Geographies of the Punjab Boundary Commission

Introduction

In the following two chapters, I return to a selection of documents from the Punjab Partition archive to find and examine the traces of geographical knowledge contained within them. In this chapter, I use the transcript hearings and memoranda submitted to the Punjab Boundary Commission (compiled and published as a series of four volumes by the government of Pakistan in 1983) in order to show in relatively broad terms how colonial forms of geographical knowledge were put to work in the various claims made to the Punjab Boundary Commission. In Chapter 5, I hone in on some of the more specific geographical work produced behind the scenes to shore up or refute different claims that were presented to the commission during the hearings. The historiographical claim that geography and geographers were not consulted adequately during the process is not untrue, but both chapters comb official and personal archives to show that geographical knowledge *was* deployed in a variety of strategic ways that had consequences for the final award.

The geography we find in the Punjab Boundary Commission documents was not always accurate, but often these inaccuracies were intentional, meant to obscure or unsettle the possible conclusions that might be made from the data in order to create a set of alternative territorial claims for consideration. I argue in this chapter that, in order to understand the geographical processes at play in

Mapping Partition: Politics, Territory and the End of Empire in India and Pakistan, First Edition. Hannah Fitzpatrick.

the partitioning process, we should think of geographical knowledge not simply as a neutral or objective science that was badly distorted or manipulated for political ends but as having its own genealogy rooted in imperial ideologies of territorial conquest and administration and colonial modes of counting and categorisation. Geographical knowledge and expertise were not politically neutral, and continuing to appeal to its possible objectivity nearly 75 years after Partition relies itself on the ideology of scientific objectivity that Partition *could* have been objective at all. This is not to say that the geography could not have been done more accurately. It certainly could have. But there were very serious limits to the possibility of geographical data and geographical techniques to resolve a political conundrum for which there was no acceptable territorial solution.

The chapter begins with an overview of the structure, membership and remit of the Punjab and Bengal Boundary Commissions, outlining how they were first and foremost judicial committees, rather than technical geographical committees. This is then followed by a consideration of the geographical terminology used in the terms of reference for the boundary commissions, as well as analysis of some of the ways that the terms of reference were challenged by various parties during the Punjab Boundary Commission hearings themselves. These sections provide the foundations for the final section, in which I conduct a critical cartographic analysis of key maps and arguments presented to the Punjab Boundary Commission.

The Punjab and Bengal Boundary Commissions

In his 3 June announcement over All India Radio, Mountbatten proclaimed that in the event of an Indian vote for Partition in the Punjab and Bengal, 'a Boundary Commission be set up by the Governor-General, the membership and terms of reference of which will be settled in consultation with those concerned' (Sadullah, 1983, p. vii). At Nehru's suggestion, it was decided that the boundary commissions would be made up of four Indian members of 'high judicial standing', nominated by the Indian National Congress and the Muslim League and overseen by a 'neutral' chairman (Sadullah, 1983. p. viii). While Jinnah had wanted the boundary commissions to be made up of members from the international community, preferably from the United Nations, this suggestion was vetoed by Nehru on the grounds that it would take too long, and by Mountbatten over the concern that international opinion would judge the British unfit to oversee their own decolonisation of India. All agreed to Jinnah's suggestion that one person should chair both boundary commissions, and they decided upon Sir Cyril Radcliffe, a successful and highly regarded lawyer, as the one for the job. The published documents that remain from this period, the statesmen themselves, and many historical accounts of Partition generally take for granted the fact that the boundary commissions were primarily concerned with the law. Scholars who have noticed, including Chester (2009), Khan (2007) and Chatterji (1999) among others, tend to point to this particular aspect of the boundary commission as one contributing factor in the failure of the new borders to mitigate violence and facilitate functional diplomatic relations between the two states of Pakistan and India.

This tension between law and geography in colonial boundary-making was not unique to Partition and was in fact a feature of boundary-making at this time. It was often geographers who raised the issue because geographical data was often subordinate to legal or political concerns that took precedence. For example, the Durand Line, the border between Afghanistan and British India drawn in 1893, was delimited without geographical expertise, a decision that Thomas Holdich (1916) disagreed with. John Donaldson and Alison Williams (2008), writing on boundary commissions specifically, note that, while the work of boundary commissions is often legal work, geographical ideas and practices are nevertheless deployed in the making of an official boundary. Geographical knowledge is still central to arbitrating legal border disputes even if that geography is not highlighted in explanations or summaries of the process. As I sought to demonstrate in Chapter 2, using Holdich (1916) and Jones (1945), boundary-making is both a geographical 'art' and a 'science', and geographical considerations both freight and complicate our understanding of ostensibly legal and political discourses on bordering.

Mountbatten declared on 3 June that India would be granted independence 'in accordance with the wishes of the Indian people themselves' (Sadullah, 1983, p. 4). While Chester (2009) and Khan (2007) both point to this as an insistence by the British to paint Partition as being motivated and managed by Indians, thereby ascribing responsibility to the Indian elite and absolving the British of any blame, a cartographic reading suggests that this insistence upon 'Indian responsibility' might also be rooted in a logic which was similar to a key element of colonial boundary-making as Holdich saw it: local involvement. Holdich wrote,

> In the actual, practical and troublesome business of a boundary settlement what is it that has proved to be the deciding influence on the adoption of its position apart from considerations as to its physical nature? The very first and greatest necessity is a careful and tactful inquiry as to the wishes of the people concerned. This inquiry must be local. (1916, p. 27)

And Jones (1945) reiterates the point in *Boundary-Making*, noting that local knowledge is a necessary requirement for a boundary to succeed. He argued that it should be harnessed as part of the effort to provide comprehensive and authoritative data on the needs and desires of the localities, which impacted directly on the delimitation and demarcation of borders.

Of course, the spirit of such guidance was hardly followed in the case of Partition. As Holdich continued,

> It is not for high personages sitting in solemn political conclave to decide on evidence which is always contradictory and uncertain what the people, whether nomads of the Asiatic steppes, tribesmen on the African plains, settlers on a disputed American frontier, French patriots or German apostles of Kultur, may consider to be the fulfilment of their desires on a matter which means life or death to them, whilst it is only a question of political exigency and bargain to the Government concerned. (1916, pp. 27–28)

Lucy Chester notes that the Indian nationalist elites took 'decisions. . .operating under British pressure' that 'in some cases ran counter to popular welfare' (2009, p. 1). Partition *was* decided upon, ultimately, by 'high personages sitting in solemn political conclave', and the evidence they deliberated upon has been shown to be 'contradictory and uncertain' with regard to 'what the people. . . consider[ed] to be the fulfilment of their desires'. It is unlikely given his own establishment position that Holdich meant what he said about 'life and death' in a subaltern way. But he judiciously intimated that Partition would, quite literally, 'mean life or death' for so many of the people affected by it. The geographer Oskar Spate, who acted as an independent advisor to the Muslim League in 1947, was well aware of the high stakes for people who would be caught up in the fallout of the boundary-making process.

The boundary commission hearings opened on Monday, 21 July 1947, taking place concurrently over 10 days in the High Courts of Lahore and Calcutta. The Punjab Boundary Commission received 51 official claims. Mountbatten and the nationalist leaders agreed, after much negotiation, which relevant parties and communities were deemed to have a stake in the final boundary, and would therefore be permitted to submit memoranda and testify before the commissions.

The Punjab hearings opened with the Congress Party's case, which was presented by a Parsi lawyer named M.C. Setalvad. The Muslim League case was presented by a highly regarded Ahmadi lawyer, Muhammad Zafrullah Khan, who after independence became Pakistan's first minister of foreign affairs, the first Asian person to preside over the International Court of Justice, and later served as the President of the UN General Assembly. The Sikhs, who joined the Congress to present a joint case, also presented their own case, as did the Scheduled Castes (or Dalits), the Christian community, the Anglo-Indian community and representatives of the Bikaner princely state. Each group formed a delegation to prepare cases for each boundary commission, gathering evidence from an eclectic range of sources, most of which were drawn from what Stoler (2002) has critically termed the 'colonial archive': academic texts including historical accounts, ethnologies and other surveys, district gazetteers, government reports, census data and other population statistics and maps. The written materials were collected and submitted as memoranda to accompany the oral testimonies delivered during the hearings. These materials were not kept secret, and so the parties had access to the data and reports submitted by all.

Yasmin Khan argues that the sheer volume of information submitted to the boundary commission created 'heightened expectations' about the final outcome among the Indian public (2007, p. 108). Oskar Spate remarked that even more materials were submitted to the Bengal Commission. The hearings were open to the public, although tickets were required, and the press and public galleries were crowded. The government had imposed a ban on reporting the hearings in the Punjab in an attempt to stem the tide of rumours and hearsay regarding the eventual award. This, of course, had the opposite effect. Oskar Spate spent most of his time between his hotel room, where he prepared maps and geographical arguments for the League, and the courthouse, which he described in his diaries: 'Found right court-room by observing concentration of armed police. Gothicy room, fairly cool. . . big dais with shabby royal arms; usual dust and general air of stuffy archives. No fewer than 10 peons, etc., shuffling judges' papers about at one time' (Spate, 1947).

Calcutta and Lahore were both historically and regionally significant cities in India, but the capital of British India was in New Delhi. The hearings were significant in that they took place in the political centres of the two provinces under negotiation, rather than in Delhi. This meant that many members of the political classes who were based outside of Calcutta and Lahore became hyper-mobile during this period, traveling between the two cities and Delhi to attend both sets of hearings and bringing much of the political debate (and rumours) with them. Khan writes, 'The centre of political gravity was shifting from New Delhi to the offices and front rooms of clerks, petty officials, policemen and administrators' (2007, p. 105).

Spate observed in his diaries the busy and crowded atmosphere at Zafrullah Khan's residence, where the League's team was based. Radcliffe, however, was notably absent and the commission hearings (in contrast to the final closed-door deliberations in Simla, which I discuss later) were in many ways an Indian affair, in which geographical knowledge became one tool by which territorial power and claims to sovereignty were put forward and negotiated. Importantly, many of the important geographical debates (most notably those concerning Gurdaspur and Amritsar-Tarn Taran districts, which will be discussed at length in the following chapter) seemed to have little impact on the final award.

I have written elsewhere about how the courtroom setting of the Punjab Boundary Commission was instrumental in the making of the territorial claims submitted, by rendering geographical data into evidence that the judges and lawyers on the commission could assess (Fitzpatrick, 2019). The courtroom worked in two ways. First, the courtroom produced a process that was, primarily, legal rather than geographical. The primacy of judicial expertise, the order of proceedings and the structure of arguments, the form and style of the memoranda submitted and the mobilisation of the norms and language of the law all worked to produce this legal process. Second, this judicial process framed geographical data as being both debatable and in competition with other forms of data. In a more recent yet parallel context, Eyal Weizman (2012), in his book on humanitarian violence and the 'lesser evil', writes about the legal battles between Palestinian farmers and landowners and the Israeli state over Israel's separation wall (or security fence, as the government calls it). Humanitarian lawyers often found themselves arguing on behalf of Palestinian villagers, often farmers, whose livelihoods and communities were impacted by specific contours and locations of the wall. Weizman observes that, during some of these trials, the use of a three-dimensional, scale topographic model of the wall and border regions in the courtroom 'helped arrive at a verdict on the "behaviour" of the wall itself' (2012, p. 76). Various parties drew multiple lines on the model, representing the actual location of the wall, the Green Line (the international boundary) and proposed alternatives. The legal process of deciding on the route of the wall, therefore, was shaped by the knowledge contained within and upon the model. Weizman writes, 'The details of the route are, however, not the direct product of top-down government planning. The route's folds, stretches, wrinkles and bends plotted the relative force of different participants brought to bear on it by the different parties and the relative force of their arguments' (2012, p. 77).

In the Punjab in 1947, the putative border was constructed through a similar process. To use Weizman's words, it was at once an object of debate and a means of creating a 'geographical grammar' for 'the law' to reshape physical reality (2012, p. 73). The American and British boundary-making experts Jones and Holdich were pertinent to what happened in the courtroom. But the courtroom itself, and the way it put the putative border and boundary-making practices on trial, so to speak, was integral to the way different elements of boundary-making were to be implemented. The presentation of legal arguments, bolstered by the use of cartographic and statistical evidence by the various political parties and their representatives, created the conditions whereby three potential boundaries were considered and negotiated: the notional boundary, the Muslim League line and the Congress-Sikh line. As Weizman writes of the Israeli security wall, 'It is in this context that the wall started appearing as a "political plastic" – a spatial product made and remade as political forces assume physical form, a diagram of the balance between the forces that shape it' (2012, p. 77). Similarly, the India-Pakistan border became a form of 'political plastic', shaped by competing claims and political forces. While the processes in both contexts worked to obscure the culpability of the state in creating the conditions for the border to exist in the first place, unlike the Israeli courtrooms, where the wall's material effects are debated and adjudicated, the boundary-making process in the Punjab brought into being a formal legal boundary that did not yet exist. Nevertheless, the courtroom proceedings put the legitimacy and authority of geographical data on trial.

The Terms of Reference for the Boundary Commissions

The terms of reference set out in Mountbatten's radio statement became one of the most contested and controversial elements of the boundary commissions and were never, even after the hearings, formally clarified. Perhaps because they had been agreed upon by all sides, the terms were vague and open to conflicting interpretations. They reflected a lack of technical geographical knowledge on the part of the government and the main party leaders and seemed, to Oskar Spate, specifically designed to keep the process as vague as possible. For the Punjab, the terms stated, 'The Boundary Commission is instructed to demarcate the boundaries of the two parts of the Punjab on the basis of ascertaining the contiguous majority areas of Muslims and non-Muslims. In doing, so it will also take into account other factors' (Sadullah, 1983, p. xii). Mountbatten stipulated that census data would be used to determine majority areas and a notional boundary was subsequently drawn using that data. But this notional boundary, Mountbatten continued, would almost surely differ from the final boundary decided upon by the boundary commissions, as representatives from each of the main parties would contribute to the process. This was part of the British discourse of 'Indian responsibility' highlighted by Chester (2009) and Khan (2007), where the British colonial government insisted on the central role played by the Indian political parties throughout the process. The terms of reference did not specify the unit at which contiguous majorities would be taken, nor did it specify what was meant by 'other factors'. The terms also misused Jones' term 'demarcate' (the final stage of the boundary-making process, the creation of the physical border on the ground). If one were to apply the boundary-making terminology of Jones or Holdich here, the boundary commissions were, in fact, tasked with 'delimiting' the boundary. Perhaps unsurprisingly, due to the ambiguous phrasing of the terms and Mountbatten's understanding of them, much of the debate in the Punjab Commission hearings revolved around the terms of reference themselves. Of course, the terms of reference indicate a lack of engagement on the part of the top tier of colonial and Indian leadership with geographical practice, but they also indicate the mobilisation of geographical knowledge in *some* capacity. Indeed, the terms did not, for example, state that the boundary commissions would 'determine' or 'draw' the boundary.

Chatterji (1999) and Chester (2009) might argue that this indicates the British desire for a 'veneer' (a term Chester uses, p. 3) or façade of geographical accuracy, but which in fact obscures the political motivations that disrupted what could or should have been an objective, even evidence-based, geographical exercise. As Chester argues, 'it was not the location of the Radcliffe boundary but the flawed process of Partition that caused the terrible violence of 1947' (2009, p. 1). But the broader literature on boundary-making is fraught with both philosophical and practical debates about the purpose and productive capacities of borders and boundaries. Interestingly, Rankin and Schofield (2004) highlight the tendency for boundary commissions (and boundary-making processes more generally) to lack clarity, and for such ambiguity to be linked to tension and conflict, citing the Partition of Ireland in 1920 and the Iraq-Kuwait boundary-making process in 1991–1993, as examples of such ambiguity. The application of the kinds of knowledge at work in the boundary-making literature exemplified by Holdich, Jones and others was hardly straightforward. In the Indian context, the Partition of the subcontinent was conducted partially within the discursive realms of the literature of boundary-making, but Partition also fit the broader trends, identified by border studies scholars more recently, whereby boundaries were given scientific legitimacy through boundary commissions and cartographic practice, but did not necessarily live up to the ideals of objectivity, rationality and scientific accuracy enshrined in boundary-making treatises.

Contiguity according to geographers refers to the unity of territory, where a spatial area is singular and unified. Contiguity is, of course, a defining geographical feature of the modern nation-state, and is achieved through the delimitation of clearly defined and effectively demarcated and maintained borders. Territorial contiguity based on majority populations was intended to address the primary concerns of the Indian nationalists regarding Indian independence, and the issue which lay at the heart of Jinnah's Pakistan claim: the future political position of religious minorities in a centralised government elected through a system of proportional representation. The original articulations of Pakistan, however cartographically vague, had posited that Muslim-majority provinces in their entirety should be included in Pakistan. The Congress Party was unwilling to accept this, as the Punjab and Bengal, both resource-rich provinces with urban centres and important industries and with sizeable non-Muslim populations, would have gone to Pakistan.

The Muslim League representative, Muhammad Zafrullah Khan, alluded to the geographical significance of this decision in his testimony to the boundary commission, arguing that the geographical units for determining contiguity were central to the process of creating Pakistan and to the creation of a 'workable boundary'. He said, 'Now I have submitted that so far as the division of India was concerned, a standard could have been a province but that is no longer available' (Sadullah, 1983, p. 292). Contiguous majority areas were based on simple majorities and in some cases, the majority was very small, especially when taken at smaller units, leaving relatively large numbers from minority populations. The proposed boundary maps, which were exercises in ethnographic mapping, do not show these populations, and the minorities, some of which were substantial, are completely erased from them. This was ethnographic mapping in the political interest of majority population.

These maps simplified and rarefied the ethnic and demographic complexity of the Punjab. However, the terms of reference sought to make some sort of amends for this by including a provision for what were called 'other factors'. The issue of 'other factors' became the primary focus of the Congress claim – and unsurprisingly, of course, because the Muslim League had the advantage in the Punjab on the question of contiguous majorities. The Sikhs emphasised that the 'other factors' clause had been put in to enable them to represent their interests, especially with regard to sacred sites and places of worship, while the Congress delegation insisted that the 'other factors' clause included economic and infrastructural concerns. The Congress team argued that the 'other factors' clause should be given more weight in the final award, and attempted to use these 'other factors' to show the economic and cultural unity of the disputed area of the Punjab. Oskar Spate disagreed, writing in his notes that the 'other factors' clause was included primarily to take into account challenges discovered during the actual demarcation phase, rather than to inform argument and debate during the delimitation process.

Challenges to the Terms of Reference

Concerns of Infrastructure

The Partition of the Punjab presented a serious cartographic conundrum, where majority populations crossed economic, political, religious, and environmental lines. We have already seen that for Jones, the technical aspects of boundary-making were of limited use when it came to dealing with this particular kind of population issue, and that he surmised that in particularly difficult situations, population transfers might be the only solution. Survey of India maps depict in great detail both the diversity of the population and the unity of the infrastructure in the Punjab. In addition to population distribution, they show the myriad sacred sites and religious centres, rivers, canals and railways that connected and organised the province. The 1931 gazetteer atlas map (Figure 4.1) depicts the railways, rivers and canals that ran through the contested central belt of the province. The Punjab Boundary Commission had essentially been tasked with dismantling the Raj's most administratively unified and developed state. As the last region of India annexed by the British, the Punjab had been the site of a number of large-scale colonial development projects, including roads, a more developed railway network and an extensive canal system implemented to increase agricultural productivity. The canal system was fundamental in the creation of a modern colonial geography in Punjab. It created a more unified system of waterways that could be centrally managed and allowed for a significant transfer of the Punjabi population through a program of land grants in newly fertile regions in western parts of the province. As Tan Tai Yong has shown, many Sikhs who had served in the First World War took advantage of government land grants after returning to the Punjab from Europe in 1919 (2005).

Radcliffe himself was especially concerned with the allocation of water resources and focused attention on the canals. Meanwhile, as I will show in more

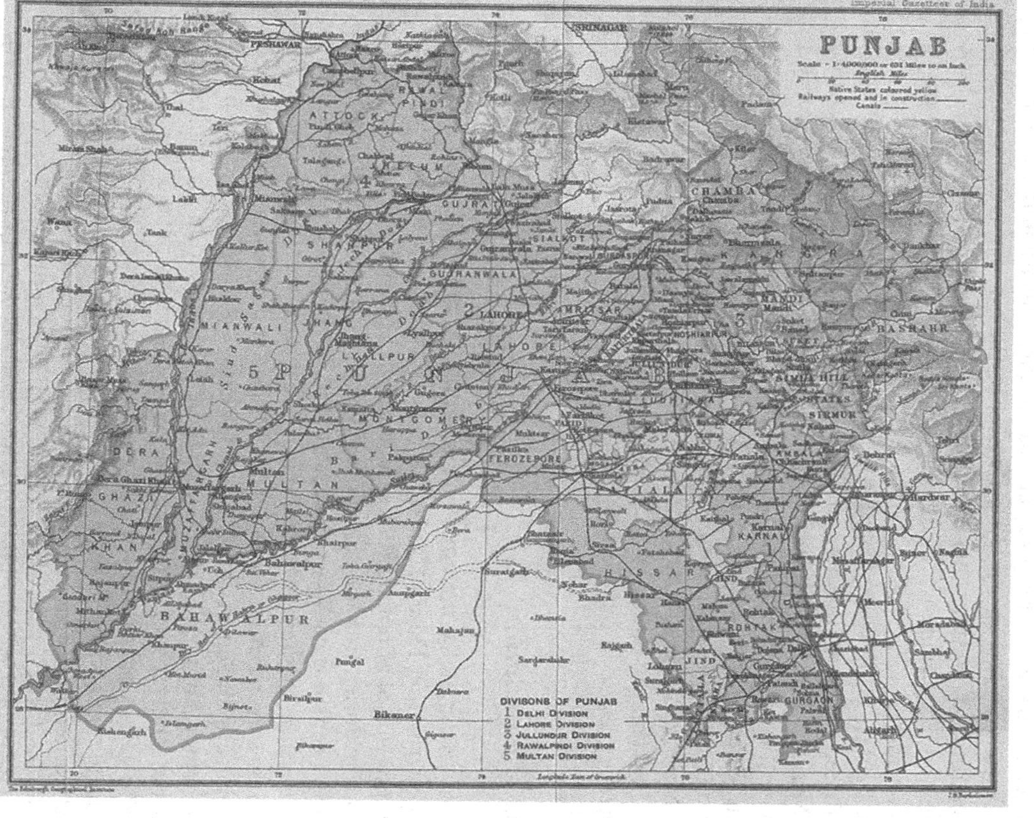

Figure 4.1 Map of the Punjab, 1931. Map showing areas under direct British administration and Indian princely states. Numbers correspond to the five divisions of the Punjab. Railway network, including both completed lines and lines under construction, represented by dashed lines. Canal network represented by solid lines marked with hashes. The disputed central belt of the Punjab encompassed the Lahore Division (2) and the Jullundur Division (3). The railways and canals were designed in such a way as to connect and unify, rather than divide, the province.

Source: The Imperial Gazetteer of India Atlas.(Hunter et al., 1908, 1931) / Clarendon Press.

detail in the next chapter, Spate's maps and diaries display his keen interest in the problems and possibilities of the railway networks. He simultaneously understood them as conduits for mobility, possible military fronts, and as one possible way of facilitating corridors or enclaves. Both the League and Congress stipulated in their arguments that a border which criss-crossed railway lines would be unacceptable. How and where, for example, would cross-border checks be conducted? How would responsibility for maintenance and security be allocated? (Sadullah, 1983). The busiest and most important sections of the railway network in the Punjab were located in the Lahore-Amritsar and Delhi areas. The Muslim League drew on Spate's geographical knowledge here to bolster their claim in order to maximise resource access via the railways and canal system. They connected the issue of canal headworks to their wider arguments about territorial continuity, arguing that the political ramifications of having headworks in one country while water flowed through the other country were too serious to consider.

To make matters more complicated, this issue around canals and water was in tension with the primary objective of the boundary commissions: to partition based on Muslim and non-Muslim-majority populations. As one example to illustrate how this worked, Pathankot was a highly contested region during the negotiations. Located in northern central Punjab and bordering the district of Gurdaspur and the princely state of Kashmir and Gurdaspur, Pathankot was the site of the Madhopur Headworks on the Ravi River. Pathankot was eventually awarded to India on the basis of population, while most of the River Ravi flowed through territory awarded to Pakistan.

Concerns of Population

The terms of reference for the boundary commission ran roughshod over these issues of infrastructure. Despite the 'other factors' clause, the terms centred and prioritised the issue of population. The information upon which the adjudication of contiguous majority area was taken came primarily from the 1941 and sometimes the 1931 census data. The census was used by the two delegations to draw maps for reference in their memoranda to the boundary commissions. Some population maps from before 1940 were published and available, the most accurate being those included in the *Imperial Gazetteer of India* Atlas, which I discussed in Chapter 2 in relation to the role these maps played in the construction of an imperial hierarchy of geographical and population data. Here, I return briefly to the gazetteer maps once again to show how they, and the bordering imaginations they engendered, were mobilised for the purposes of Partition.

The first edition of the gazetteer to include a separate atlas volume was published in 1909 and was updated in 1931. The source maps for the atlas were originally drawn by the Survey of India office but were prepared in Edinburgh by the well-known family firm run by J.G. Bartholomew. These maps depicted much of the geographical information contained within the volumes of *The Imperial Gazetteer of India* in cartographic form. They depict geological and climate data; population data showing the breakdown of the Indian population based on such categories as race, religion and language; infrastructural and administrative data (including canal routes, railway lines, military divisions and economic resources) and historical maps. All of this was designed primarily to illustrate the expansion and consolidation of British power across the subcontinent, but in 1947, these cartographic representations of British India were put to work in a new way: to expedite decolonisation.

In looking at the gazetteer maps, we see a colonial cartographic representation that looks strikingly similar to later maps of Pakistan. The map shows the regions that were predominantly Hindu, Muslim, Christian, Buddhist, Sikh and Animist (which the colonial government used to refer to tribal populations). Shown in green, the Muslim-majority areas dominate the north-western region of the subcontinent and occupy a sizeable chunk (Bengal) in the north-eastern corner, while a very small Sikh pocket sits in a corner of the Punjab where the Hindu- and Muslim-majority regions meet. Similarly, the ethnographic and linguistic maps depict the spatial distributions of the various social and racial categories that the British identified among their Indian subject population. As mentioned above regarding the maps submitted to the boundary commission specifically, the immediate effect of such ethnographic mapping was the obfuscation of local differences and variation. This kind of mapping emphasised territorial boundaries which reflected the social boundaries that the British deemed salient to the political and cultural organisation of India.

Choudhary Rahmat Ali, introduced in the previous chapter, wrote specifically of this 'British map', calling upon its authority to argue that Muslims had a claim to sovereign Indian territory, based on a combination of historical ties to place and concentration of population. Similarly, when Jinnah said, in his 1940 presidential address to the Muslim League, 'we find that even according to the British map of India, we occupy large parts of this country where the Mussalmans are in a majority, such as Bengal, Punjab, NWFP, Sind, and Baluchistan', he was invoking these cartographic images of colonial India in order to claim that the social boundary between Hindus and Muslims was reflected in a clearly visible spatial boundary that had long been depicted on the 'British map' (1940, in McDermott et al., 2014, p. 503). In the Lahore Resolution, this social boundary was both reified and rearticulated as a political boundary, and the 'British map' became a cartographic justification for claiming territorial sovereignty for India's Muslim population. Meanwhile, the Congress mainstream preferred to read the map as a cartographic rendering of India's ability to maintain its political unity and overcome challenges associated with its cultural diversity (Figure 4.2).

Ethnographic mapping had served its purposes for the Raj. In Chapter 2, I discussed the processes by which colonial forms of knowledge gave rise to

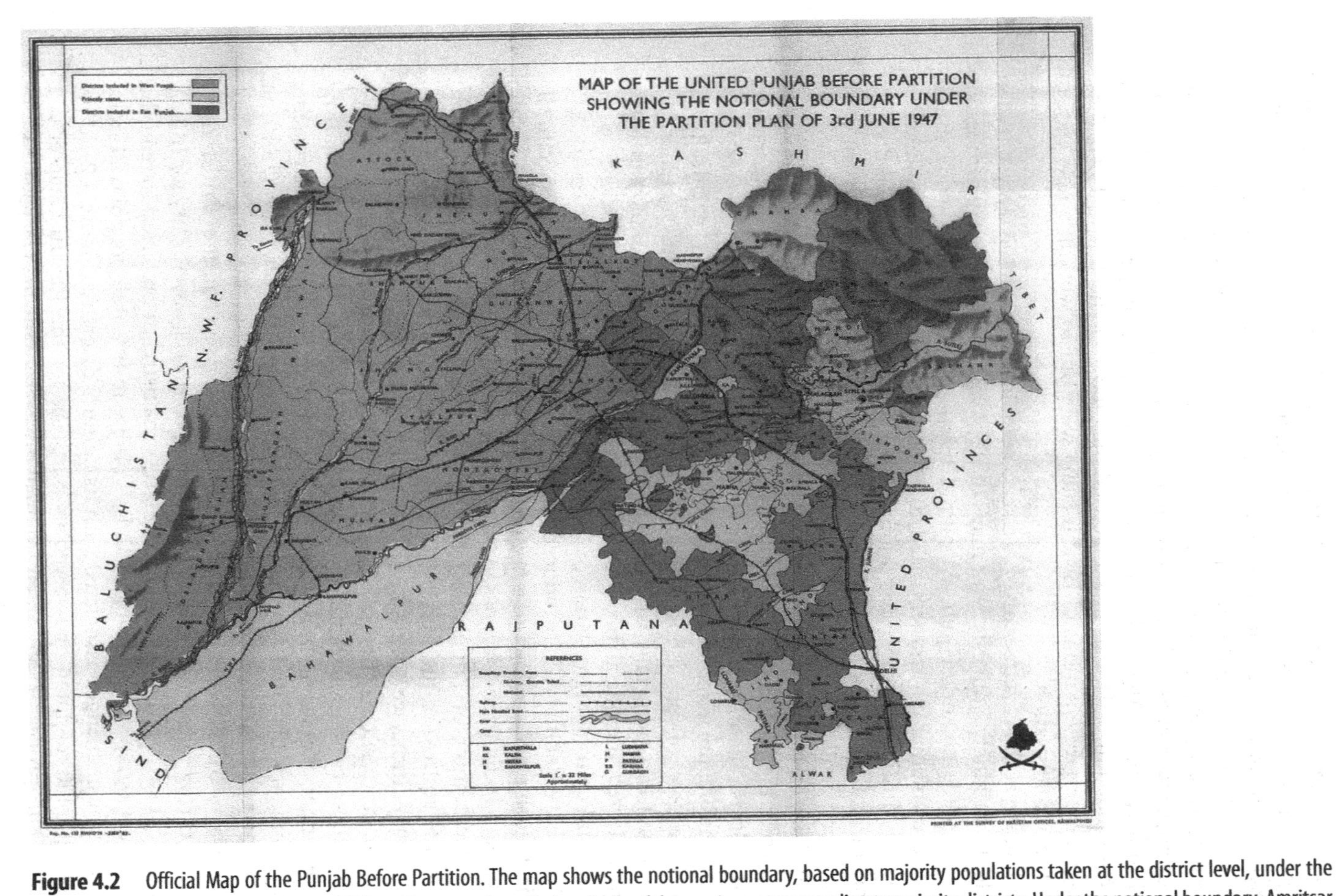

Figure 4.2 Official Map of the Punjab Before Partition. The map shows the notional boundary, based on majority populations taken at the district level, under the Partition plan of 3 June 1947. The notional boundary runs through the middle of the province, corresponding to majority districts. Under the notional boundary, Amritsar went to India and Gurdaspur went to Pakistan.

Source: Panjab Digital Library/http://panjabdigilib.org/webuser/Download//last accessed 23, November, 2023.

particular knowledge categories and strategies for governing the Indian population. In Chapter 2, I illustrated some of the ways in which ethnographic mapping had become incorporated into the literature and practice of boundary-making after the First World War, exemplified especially by the work of Stephen B. Jones. But after World War II, ethnographic mapping became a more refined tool for forging new territorial nation-states. In the Indian context, of course, the boundary commissions did not base their boundaries on the same kinds of racial categories as the Americans and French had sought to do in Paris in 1919. In India, it was the category of religion, rather than race, which was taken to be the foundation upon which the boundaries of new nation-states would be drawn. However, Javed Majeed notes that 'race' in the Indian context was more fluid than it was in the European context. In India, religion, language, tribe and caste all fed into the classification of difference and configuration of race (Majeed, 2009). However, the boundary commission maps, and the historicisation and essentialisation of differences which attended British and Indian nationalist discourse after 1940, do reveal some remarkable similarities to the 1919 European context. The British government attempted, through the publication of an atlas volume to accompany the *Imperial Gazetteer of India* after 1901, to depict on maps the socio-biological and cultural categories it had spent decades attempting to identify amongst its subject population in India (Hunter and Bartholomew, 1908).

A territorial partition was, essentially, a geographical practice designed to create territorial boundaries which spatialised these understandings and categories. We know that Partition in India was a convoluted political process, and one that cannot be reduced to an attempt to simply divide territory along communal lines; indeed, much of the revisionist historiography on Partition has been designed to specifically challenge that assumption. Recalling Gyanendra Pandey (2012), the construction of communal identities in India in the 19th century was intertwined with the development and hardening of colonial categories. However, I have attempted to show how colonial discourses of the Indian past and the Indian people constructed the discursive arena within which Indian nationalism developed.

The arguments and maps presented to the Punjab Boundary Commission illustrate the various and sometimes contradictory ways that the different Indian nationalist organisations understood and articulated the geography and territoriality of the Punjab. They mobilised these competing geographical and territorial visions in order to present convincing narratives of the future geographies of India and Pakistan. The maps display the impossibility of reconciling the multiple narratives of territorialised nationalism with the realities of technical boundary-making. They also show the limits of the map as a reliable form of evidence, using cartographic authority and scientific geographical data to communicate cultural and religious narratives and aims. Time also plays a role in how we read and interpret both the archived maps from 1947 and the border. Before 14 August 1947, the boundaries existed in a kind of liminality, all possibilities without legal or official certainty. After the announcement of the Radcliffe Award, a single boundary became the reality, thereby inscribing upon the territory and the map a narrative, cartographic and territorial certainty that shaped both the future of the new nation-states and the new narratives of the subcontinent before Partition. This explains why a colonial map from 1909 can provide a powerful and popular cartographic representation of a geopolitical event that took place three decades later.

Reading the Maps of the Punjab Boundary Commission

In this section, I introduce the key maps detailing the broad territorial claims produced for the Punjab Boundary Commission by the prominent delegations represented in the hearings. I examine these maps alongside some of the testimony that was delivered in order to contextualise them, as well as some of the debate they generated. I do so in order to show some of the ways, both subtle and overt, that geographical rhetoric and data were mobilised as part of the wider legal process. This discussion provides both background and context for the analysis in the following chapter of some of the more granular, experimental or messy draft maps that were never published or even officially presented to the boundary commission. This section therefore analyses some of the more fundamental geographical arguments that were introduced during the process, while the next chapter looks more closely at the ways that such arguments could be both bolstered or refuted by the very geographical concepts and principles they appealed to.

The terms of reference dictated that the arguments demonstrate 'contiguous majority areas' of the contested regions, which, at first glance, seemed relatively easy to determine. According to the colonial state's archive of itself, there already existed a glut of population data collected via the seemingly systematic census over 70-odd years, as well as a long-established survey office that had provided accurate maps of India for over a century. The boundary-making process should have been, according to this idealised narrative the state told of itself, a matter of simple calculation and objective mapping. The reality of course was that the source data was far more problematic, open to multiple interpretations and manipulations.

Despite the large area of land subject to Partition, it was certain smaller areas, some as small as a *tehsil* or a town, that were the most contested during the process. Each of the contested areas was claimed using variations on geographical themes, throwing multiple geographical justifications into the mix in the hope that the justices would find it convincing. The Muslim League's claim to the Punjab was for the most part based on majority Muslim areas calculated by taking the *tehsil* as unit, which according to Spate was a sound geographical technique (putting aside the generally acknowledged issue of the reliability of the census data itself). The League's claim to contiguity rested on the argument that one of the most important decisions facing the boundary commission was the question of which geographical unit should be used to calculated population distribution. The terms of reference did not specify this, but the calculations based on different units would change the course of the boundary. While many of the conversations and debates in the hearing transcripts refer to the larger and more commonly referred to administrative unit of the district, the League argued and Spate agreed that the *tehsil* was the optimal unit for calculating majority populations, because the larger unit (the district) was not granular enough while the smaller units (*zails* and *thanas*) were too intermixed and did not clearly show any contiguity whatsoever. However, much of the debate revolved around districts, and so I use the district unit, rather than the *tehsil*, when outlining why some areas were hotly contested and how the geography of the final award enabled certain outcomes. The issue of scale was therefore central to the entwined problems of majority and contiguity.

In contrast, the Congress argument focused on undermining trust in the 1941 census data and emphasising the 'other factors' clause in order to obfuscate and challenge the League's population-based argument. Their map (Figure 4.3) equally served to obfuscate the League's cartographic representation of the population, and it did this by purposefully not using a standard unit in its calculations. This was a clear demonstration of geographical rule-breaking. The map, generally referred to in the historiography as the 'Congress red map', after its strategic use of red shading, claims as non-Muslim a significant area of Muslim-majority territory by locating, for example, a non-Muslim-majority village surrounded by Muslim-majority villages. The Congress argument attempted to claim 'contiguity' based on whichever units provided a majority.

The justices on the boundary commission noticed, of course, and, when challenged on it, the Congress representative, Setalvad, had to admit that a standard unit had not, in fact, been used. The League argued that consistency in units, as well as the decision-making around which unit to adopt, was a necessary component of the boundary-making process (and a professional geographer would also have argued this, which Spate in fact did). If a standard unit was not adopted, Zafrullah Khan argued, the original Muslim League claim, which was that the entirety of Muslim-majority provinces should be included in Pakistan, would stand, and the Partition of Punjab Province itself would be called into question, given that the province itself was a contiguous Muslim-majority area. The boundary commission, Zafrullah Khan argued, 'must set themselves some standard, otherwise they will be faced with this position that in one sense, as I have already said, the whole province is a majority Muslim area' (Sadullah, 1983, p. 286).

The League's map (Figure 4.4) was based on recognised geographical principles, which was made easier of course by the fact that they had the stronger claim in the Punjab according to the terms of reference. But the Congress map

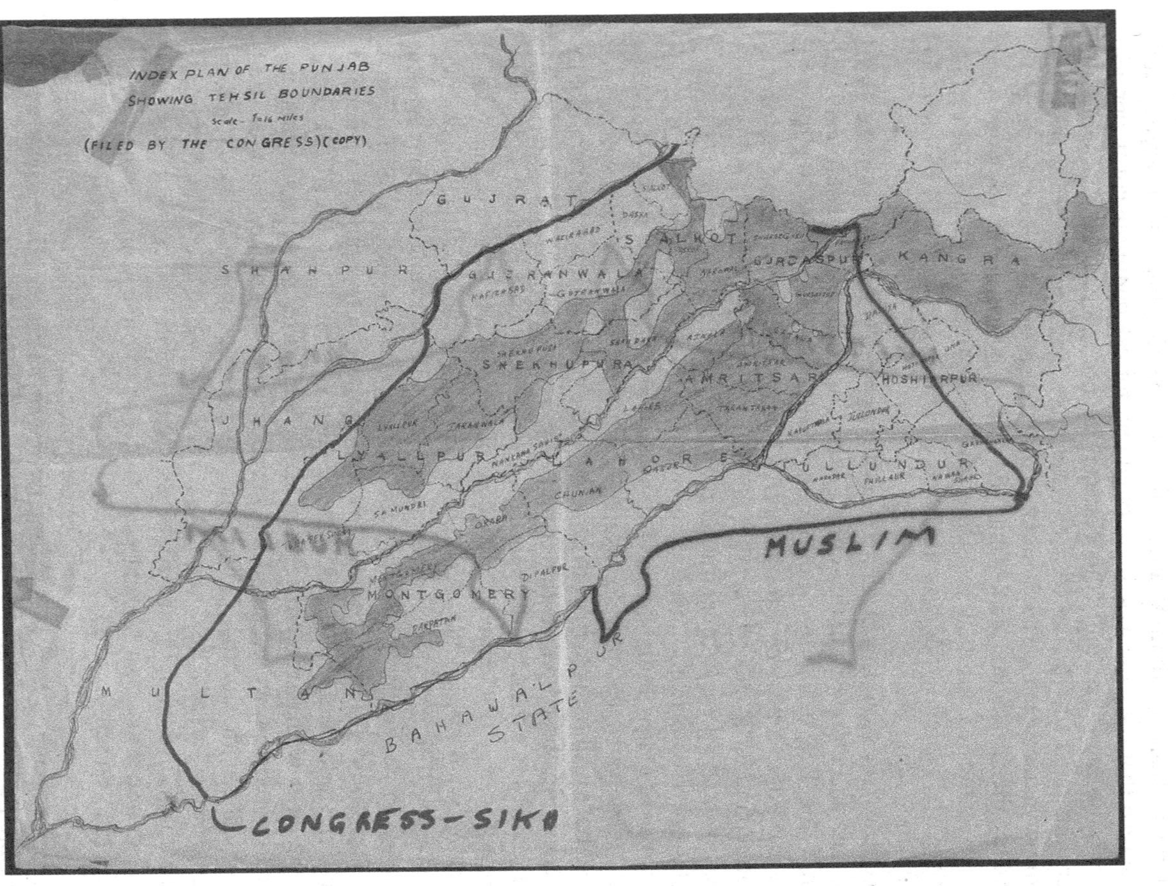

Figure 4.3 The Congress 'Red Map'. Showing the Congress-Sikh and Muslim League proposed boundaries. Non-Muslim majorities are shaded, while Muslim majorities are left unshaded.

Source: The Papers of Oskar Spate, MS 7886, National Library of Australia.

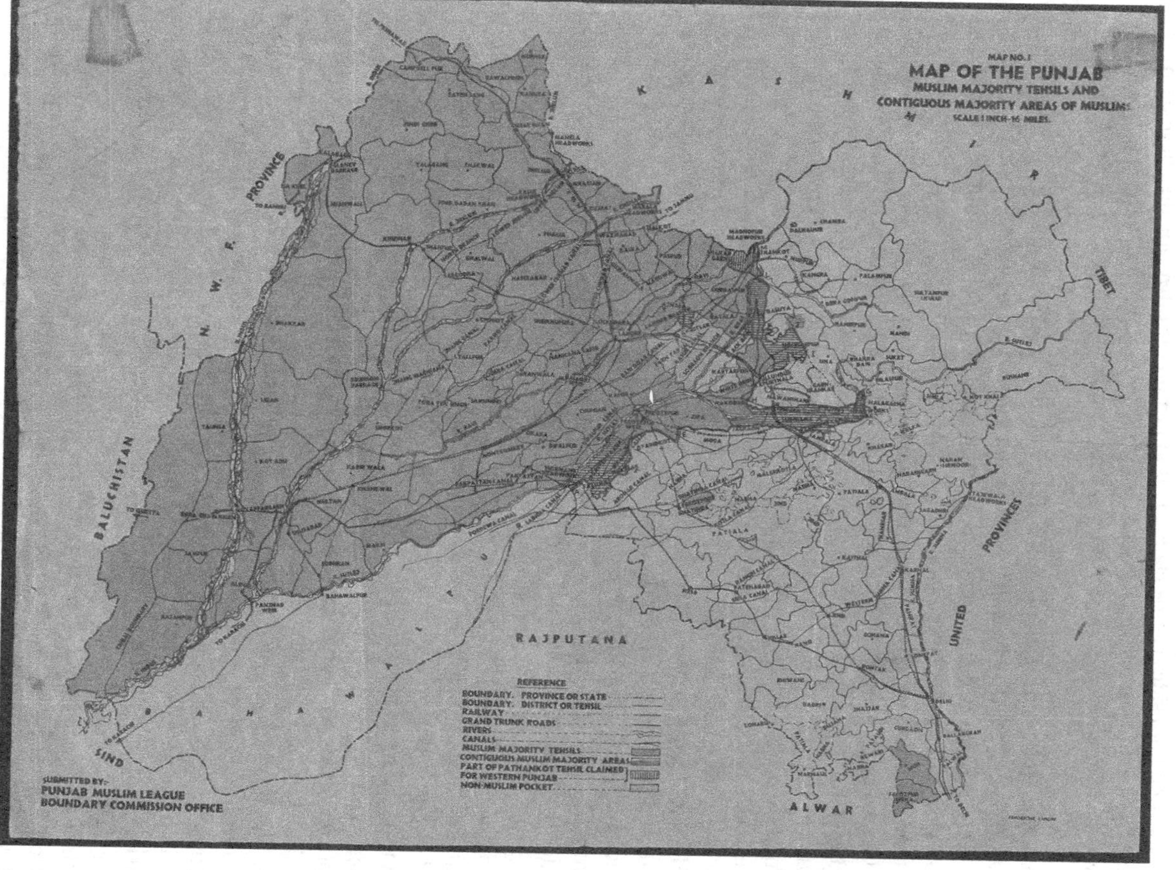

Figure 4.4 The Muslim League Map. The Muslim League map, submitted to the Punjab Boundary Commission, showing the League's proposed line, Muslim-majority *tehsils* are shaded, a non-Muslim-majority pocket in Amritsar, and non-Muslim-majority areas unshaded.

Source: The Papers of Oskar Spate, MS 7886, National Library of Australia.

represented a significant attempt at gerrymandering, and Zafrullah Khan noted that without a standard unit, both sides could claim large tracts of territory where their populations were, in fact, a minority. He argued that Congress had done exactly this with their red map saying,

> I have tried to make it clear the commission should adopt some standard, otherwise large areas on both sides can be claimed, as in fact on the other side they have been claimed, as majority areas which even a more cursory examination of the map would show were not majority areas of that particular community. (Sadullah, 1983, p. 291)

In arguing this point, Zafrullah Khan was drawing attention to the fact that the delegations were putting forward competing cartographic claims, but that those claims were not equivalent in geographical terms, emphasising that the League's position was more defensible according to the geography. The League went on to argue that not all geographical units were equivalent in this case and that the *tehsil* should be taken as the standard unit in all calculations. Zafrullah Khan similarly argued that the *tehsil* was the most accurate unit in terms of its representation of population statistics, and therefore most easily facilitated a reorganisation of administrative territory.

The Sikh justice Mehr Chand Mahajan noted one of the central issues around visualising population at different scales, which he called population 'intermixing'. He said, 'Supposing we apply the yardstick of a *zail* or a *tehsil*. We then arrive at a substantial area which on that yardstick is so intermixed that we cannot work out the contiguity' (Sadullah, 1983, p. 287). Zafrullah Khan countered, 'If you continue on the basis of a division or a district or a *tehsil* – may be that you may consider the boundary drawn on the basis of those standards undesirable or subject to modification for one reason or the other, but I do not think that intertwining of areas will arise. If you go below that standard you may fall in that difficulty' (Sadullah, 1983, p. 287). In other words, the smaller the scale, the more intermixed the population would be, and the more difficult it would be to discern contiguity, as so many localities in the contested areas of the Punjab were not homogeneous. This discussion between Zafrullah Khan and Mahajan highlights one of the limits for geographical boundary-making techniques to solve one of the political problems at the heart of the Pakistan demand: democratic representation based on geographical units in a society that had for a number of decades been administered on the basis of political, social and religious difference. The *tehsil* may have been the most accurate geographical unit for the purposes of calculating majority population, but it did not solve the intractable problem of the minority populations who were still there, whether or not they appeared as a colour shaded on the map.

Zafrullah Khan also appealed to the continuity and accessibility (not to mention authority) of the current administrative map of the Punjab. The *tehsil* was a widely used and acknowledged geographical unit already, and had been in use relatively consistently over time, and while they were redrawn occasionally, they were, compared to *zails* and *thanas* (both smaller units), less 'liable to alteration' (Sadullah, 1983, p. 293). He noted that if units any smaller than the *tehsil* were used, the boundary would follow 'a very crazy pattern' (Sadullah, 1983, p. 296). This would constitute a failure on the part of the boundary commission to create a 'workable boundary', which according to the Muslim League was expressed using the geographical understanding of spatial scale and population distribution (Sadullah, 1983, p. 291).

Despite their insistence on the need for a standard base unit, there was wiggle room; the League's proposed boundary did not always neatly follow the pre-existing borders of *tehsils*. They argued that contiguous Muslim-majority areas extended into *tehsils* that had overall non-Muslim majorities when the smaller Muslim areas were contiguous with neighbouring *tehsils*. These included, for example, certain parts of Fazilka and Muktsar *tehsils*, which the League argued 'were in continuation of the Ferozepore *tehsil*'. On the map, these areas appear as green horizontal lines. The League argued that these claims were based on a true representation of contiguity, in contrast with the Congress claims because the areas that the League claimed in the Fazilka and Muktsar *tehsils* actually had significant Muslim majorities, which were contiguous to the rest of the League's claim. The Congress map, on the other hand, attempted to use small areas of non-Muslim-majority in order to claim larger surrounding areas with Muslim majorities, which did not accurately depict contiguity:

> There is no contiguity to a compact area there in the case of the other side. Take Batala *tehsil*. Bits are left out at one end, bits are left out in the middle and other bits are taken. There is this distinction between the areas claimed by the particular method adopted by the other side and the areas claimed by the method adopted by this side. The areas claimed by us are contiguous and compact and have a majority. (Sadullah, 1983, p. 303)

The 'other factors' clause proved both useful and problematic for both parties. The League, for example, used the 'other factors' clause in order to maintain continuity in the canal system. The League map claims part of the Pathankot *tehsil*, where the Madhopur headworks were located. The canal controlled by the headworks, the Upper Bari Doab canal, irrigated the areas claimed by the League based on the key principles of the terms of reference: contiguous Muslim-majority areas. The League therefore used the 'other factors' clause to add other geographically informed claims about infrastructural integrity and security, in order to gain the most workable and cohesive award for the Muslim-majority population in what was set to become Pakistani Punjab.

Meanwhile, having the harder case to make in the Punjab, the Congress delegation tried to use the 'other factors' clause in a different way. They ultimately argued that the specific social, cultural, historical and economic context of the

Punjab gave rise to so many other factors that the province could not justifiably be partitioned along the lines determined by majority population. They claimed all of the highly contested regions on the basis of the argument that the Punjab was a single indivisible unit according to these other factors. The Congress party and the Sikh delegation both then tried to show how the non-Muslim population of the Punjab, despite being minorities in most of the contested areas, were inextricably tied to the territory. The Congress and Sikh territorial claims were almost identical, with the exception of two *tehsils* which the Sikhs claimed for India and which the Congress would let go to Pakistan. Both claimed territory extending up to the River Chenab, well into areas with a clear Muslim majority. The Sikhs staked this claim on historical and religious grounds, while the Congress justified its claim primarily through appeals to property rights and economic hardship while honing a portion of its argument on the unreliability of the census. The Sikh claim took on board numerous historical and religious sites, including most gurudwaras (Sikh sites of worship and gathering) and places connected with the gurus' lives and teachings. The Sikh memoranda included a list of all the sites they asked for – sites which were scattered throughout the contested region. Meanwhile, the Congress claim attempted to show that Hindus would experience drastic and unnecessary economic hardship as a result of forced migration from west to east Punjab, and that Hindus were more entrenched in the Punjab economy than Muslims. This was true in many ways, primarily because of the patterns of colonial favour and privilege.

The problem with such a claim was that it relied on the Muslim population not having such ties. The Congress testimony, then, presents a problematic narrative in which Muslims are often represented as transient, migratory and less economically dependent on the land. The red map reflects this overall argument by trying to show a territorial connection between the non-Muslim population of the Punjab while erasing the Muslim-majority population entirely. While the League map shows the non-Muslim population in yellow, the red map leaves the Muslim-majority areas blank, without marking their boundaries or shading them. By obscuring the Muslim population, the map provides a cartographic rendering of the overall Congress claim, which relied on a narrative of cultural 'homogeneity' and 'unity' (Sadullah, 1983, p. 69). Spate argued that according to geographical logic, the map represented an inaccurate claim to contiguity. There was an underlying logic to the red map, however, which served to obscure the geographical authority of the League's population map and the internal boundaries that it implied, supporting the Congress argument that the Punjab was interconnected and unified.

Two of the most highly contested districts were Amritsar and Gurdaspur districts. Amritsar was the spiritual home of the Sikhs and had the highest population of that community, and was also a significant urban centre for Hindus and Muslims. Gurdaspur, meanwhile, bordered the princely state of Kashmir, and whichever new state had Gurdaspur would have a claim to territorial contiguity with Kashmir, thereby strengthening its overall claim to Kashmir. The main parties (and Spate) understood the geopolitical significance of Gurdaspur at the time, and the implications of Radcliffe's eventual decision about the district reverberate today. In his oral arguments, Setalvad on behalf of the Congress claimed the entirety of the Amritsar and Gurdaspur districts for India, arguing, 'the district of Gurdaspur, considering its physical features and the topography and its situation in reference to the canal scheme, is really one unit as it were with the Amritsar district. . . this division has been made for administrative purposes but in fact the districts are one homogeneous unit' (Sadullah, 1983, p. 69). He continued, 'You will also notice that these two districts, Amritsar and Gurdaspur, are united by very close trade relations. . . there is rail and road communication coming up from Lahore to Amritsar, then from Amritsar on to Gurdaspur district and vice versa, inter-linking Kangra Valley and the district of Kangra itself with Amritsar and Lahore' (Sadullah, 1983, p. 69). Using this argument about transportation and communication infrastructure, Setalvad then extended the claim to include Lahore (an undisputed Muslim-majority city), highlighting also Lahore's location along the same canal that also irrigated Amritsar and Gurdaspur. He brought the red map into his testimony in order to make this point cartographically:

> If you turn to the map, you will notice that the headworks of the Upper Bari Doab canal are situated at Madhopur. This canal irrigates parts of Gurdaspur, a large part of Amritsar and goes on to irrigate parts of the Lahore district as well. If that tract is taken as a whole, you will find it ranges from Pathankot and extends into Lahore district and is a non-Muslim majority area and it is also contiguous. Taken as a whole, it starts from the Gurdaspur district, embraces about two-thirds of the Amritsar district and further includes more than half of the Lahore district. All this area is united by a very powerful factor, as it is irrigated by the same canal system. (Sadullah, 1983, p. 71)

The Congress claims were often constructed in similar terms: the geographical area is identified in vague terms like 'parts of', 'two-thirds' and 'more than half', and includes some of each district but not the whole of either one. Internal boundaries are elided in this narrative, where geographical units are secondary to the injunction, 'taken as a whole'.

All of the maps presented to the boundary commission are very effective at erasing the spectre of the 'oppressed minority', which haunted so much of nationalist debate. Mapping contiguity meant painting majority areas a single colour, while the reality on the ground was that the Indian population was diverse and the distribution of different groups was complex. Indeed, this very diversity was central to the Congress' rhetoric of national unity amidst cultural diversity. Historians have pointed out the irony of Jinnah's justification for Pakistan in these terms: while Pakistan was designed as a 'safeguard', that is, a political and territorial tool to transform Muslims from a minority in India to a majority in Pakistan, the Muslim League did not effectively imagine or ensure those 'safeguards' for future non-Muslim minorities in Pakistan before 1947. Indeed, by mobilising a pro-Muslim nationalist sentiment that was powerful enough to counter Congress' claim to speak for all Indians, the League made it even more difficult to promote a parallel rhetoric that included protections for minorities. But the Congress' claims of Punjabi unity and homogeneity had a similar effect of erasing the complexities and diversity of the culture and geography of the Punjab in favour of a rhetoric of homogeneity and unity.

Indian nationalists never worked out a way to deal effectively with the politico-territorial identity that is so integral to the citizens of the democratic nation-state. By appealing to the Muslim *umma* (or community), Iqbal tried to de-territorialise Indian Muslims in an attempt to create a religio-social identity at both the individual and collective levels but found himself falling back on territorial claims in order to argue for self-sovereignty for those who claimed that same religio-social identity. Jinnah backed the notion of a loose federation of Indian states, but could not produce a suitably unified and mobilised political identity for the future citizens of that federation. The Congress Party, so attached to its opposition to reserved seats and affirmative action, as well as its insistence on a powerful central government, could not devise a form of secularism that was palatable to Jinnah and the League. It is unsurprising, therefore, that the so-called battle of the maps which characterised much of the Punjab Boundary Commission's deliberations showed an inability to recognise and work through the problems of the territorial nation-state model for India's minority populations.

While the written and oral arguments presented to the Punjab Boundary Commission do often present data to show the Muslim and non-Muslim populations in many parts of the Punjab, the maps themselves do not include this data. Nor do they use any form of shading or colour coding to indicate relative majorities and minorities, a technique used in other instances of ethnographic mapping, such as Emmanuel de Martonne's 1920 ethnographic maps of Greater Romania (Palsky, 2002). The effect of deficit in the Punjab case was to erase any cartographic trace of minorities and to depict the homogeneity and unity that characterised much of the rhetoric of the idealised new nation-states as cartographic realities. So, for example, the League map showed contiguous non-Muslim-majority areas in yellow and contiguous Muslim-majority areas in green but obscured the fact that the population of the Punjab was quite mixed at the village and *tehsil* levels in the key regions at stake in the discussions. As a result, the map did not illustrate the potential for large numbers of Hindus and Sikhs (and others, including Parsis) to find themselves in the new state of Pakistan. Similarly, and perhaps more egregiously given that the foundations for claims of contiguity were more geographically and statistically dubious, the Sikh map showed the Congress-Sikh proposed boundary, with all *tehsils* on the Indian side shaded yellow, and all *tehsils* on the Pakistani side shaded green, despite the fact that much of this area was, in fact, inhabited by Muslims. Little wonder, then, that the new nation-states were not prepared for the refugee crisis which engulfed them after Partition as substantial numbers of minorities moved across the new border, effectively following the rules set out by the cartographic ideal depicted on these maps: Muslims on one side, Hindus and Sikhs on the other. Like the revenue survey maps that Michael (2007) discusses at length, the Partition maps represent a particular image of the administrative state which did not yet exist in reality, and which, once enacted, would manipulate particular patterns of mobility and identity on the ground.

Conclusion

This chapter has shown how the courtroom produced evidence of a particular kind, transforming geographical data into rhetorical and legal arguments that could be disputed and negotiated. This data often had to do with population distribution and infrastructure. Of course, more was at stake here than even population, railways and canals. The next chapter will show in more detail how religious and historical sites, especially important to the Sikhs whose most sacred place was about to be partitioned, became a significant component of the Congress and Sikh arguments. Strategy and defence were also important for the League, as well as for Spate, as we will see in the analysis of his work in the following chapter. Private property became a central argument for the Congress delegation, although as Spate noted, property was not considered a reasonable claim according to boundary-making precedent. And, finally, the Punjab boundary dispute was highly significant in the ways it created opportunities for competing territorial claims to Kashmir (the true significance of which was only learned after the fact). All of this illustrates the tensions, even paradoxes, inherent in the creation of a delimited and demarcated territorial nation-state out of the imagined nations that existed concurrently in the rhetoric and practice of Indian nationalism before independence, and in the construction of colonial categories and its corresponding cartography.

The transition from territorial potentiality to territorial reality was evident in the tensions between technical cartographic practice and imperial and nationalist imagined geographies. The next chapter hones in on this tension by examining in detail Oskar Spate's work as a technical advisor to the Muslim League and the Ahmadiyya community. Some of the maps and materials analysed in the next section made relatively little impact on the final boundary, while others played a more prominent role in the hearings and were reflected in Radcliffe's award. Spate's position was therefore an odd one, adjacent to the official process without any decision-making authority whatsoever, but intervening in small ways to affect the contours of the arguments and territorial claims presented to the commission.

References

Chatterji, J. (1999). The Fashioning of a Frontier: The Radcliffe Line and Bengal's Border Landscape, 1947. *Modern Asian Studies* 33, 185–242.

Chester, L. (2009). *Borders and Conflict in South Asia: The Radcliffe Boundary Commission and the Partition of Punjab*. Manchester: Manchester University Press.

Donaldson, J. and Williams, A. (2008). Delimitation and Demarcation: Analysing the Legacy of Stephen B. Jones's Boundary-Making. *Geopolitics* 13 (4): 676–700. https://doi.org/10.1080/14650040802275503.

Fitzpatrick, H. (2019). The Space of the Courtroom and the Role of Geographical Evidence in the Punjab Boundary Commission Hearings, July 1947. *South Asia: Journal of South Asian Studies* 42 (1): 188–207. https://doi.org/10.1080/00856401.2019.1555795.

Holdich, T. (1916). *Political Frontiers and Boundary Making*. London: Macmillan and co. limited.

Hunter, W.W. and Bartholomew, J.G. (1908, 1931). *Imperial Gazetteer of India*. Oxford: Clarendon Press.

Jones, S.B. (1945). *Boundary-making: A Handbook for Statesmen, Treaty Editors and Boundary Commissioners*. Washington: Carnegie Endowment for International Peace.

Khan, Y. (2007). *The Great Partition: The Making of India and Pakistan*. New Haven; London: Yale University Press.

Majeed, J. (2009). *Muhammad Iqbal*. New Delhi: Routledge, Taylor & Francis Group.

McDermott, R., Gordon, L.A., Embree, A.T., et al. eds. (2014). *Sources of Indian Tradition: Modern India, Pakistan, and Bangladesh*. New York: Columbia University Press.

Michael, B. (2007). Making Territory Visible: The Revenue Surveys of Colonial South Asia. *Imago Mundi: The International Journal for the History of Cartography* 59, 78–95. https://doi.org/10.1080/03085690600997852

Palsky, G. (2002). Emmanuel de Martonne and the Ethnographical Cartography of Central Europe (1917–1920). *Imago Mundi: The International Journal for the History of Cartography* 54 (1): 111–19.

Pandey, G. (2012). *The Construction of Communalism in Colonial North India*, 3rd ed. Delhi: OUP India.

Rankin, K.J. and Schofield, R. (2004). The Troubled Historiography of Classical Boundary Terminology. IBIS Working Paper no. 41. Dublin: University College Dublin.

Sadullah, M.M. (1983). *The Partition of the Punjab, 1947: A Compilation of Official Documents*. Lahore: National Documentation Centre.

Spate, O.H.K. (1947). Papers re Punjab boundary dispute (Lahore) 1947 (6-1-1). Papers of Oskar Spate. MS 7886. National Library of Australia.

Stoler, A.L. (2002). Colonial Archives and the Arts of Governance. Archival Science 2. 87–109.

Tan, T.Y. (2005). *The Garrison State: Military, Government and Society in Colonial Punjab, 1849–1947*. New Delhi: SAGE Publications Pvt. Ltd.

Weizman, E. (2012). *The Least of All Possible Evils: Humanitarian Violence from Arendt to Gaza*. London; New York: Verso.

Chapter Five
Oskar Spate, the Muslim League and Geographical Expertise

Introduction

In this chapter, I shift to one side of these heated debates over territory, history and sovereignty that took place over the course of the Punjab Boundary Commission hearings in order to examine the partitioning process itself through the eyes of one geographer-witness: Oskar Spate, a British geographer who had been educated at Cambridge and was then lecturing at the London School of Economics. Here I develop and complicate the geopolitical and cartographic story presented to the boundary commission outlined in the previous chapter by analysing Spate's notes, maps, and academic articles relating to his work during the Punjab Boundary Commission. I focus specifically on Spate's unpublished notes, maps and correspondence held at Australia National Library, Canberra, although my analysis is refracted through interpretations of Spate's published material, including the academic articles he produced on India and Pakistan in 1943, 1947 and 1948, as well as his 1991 memoir.

My argument is two-pronged: first, I aim to show, through a discussion and analysis of Spate's technical geographical work for the Muslim League, that the practice of partition as a project of applied geography was a complex negotiation between the political and the technical. I am concerned with the ways in which a scientific discourse of objectivity, rationality and functionality was deployed (with varying degrees of success, of course), both by Spate and by the main parties

Mapping Partition: Politics, Territory and the End of Empire in India and Pakistan, First Edition. Hannah Fitzpatrick.

during the Punjab Boundary Commission hearings, in order to construct arguments and maps in support of competing claims, while the claims themselves were often deemed (by Spate, by members of the boundary commission and by scholars and commentators writing after 1947) to be irrational, subjective and ignorant of the geographic specificities and technicalities of what Spate and others called a 'workable boundary'. The second component of the argument concerns the complex role of the 'ambivalent geographer' in the partition process. I aim to show how Spate was caught up in the politics of the time, both an active player and a 'witness', at once 'in the thick of it' and a 'fly on the wall' (both his own expressions, which he used to describe his experience in the Punjab) (Spate, 1991, p. 50). This aspect of the argument taps into wider questions about the uses and limits of 'expertise', specifically in the context of the mid-20th century, as the imperial world system was transformed into a system of sovereign territorial nation-states. Spate's expertise was bounded not only by the circumstances of his position within the boundary-making process (as a technical advisor to the Muslim League, rather than as an arbiter, or technical advisor to the boundary commission itself) but also by the limits of geographical discourse at the time.

While we, as 21st-century observers, are very much aware of how such limits have shaped much of the postcolonial world order, Spate himself was working in the midst of the transfer of power, and the end of the British Empire in India. His ambivalence towards empire, his concerns for the Indians and Pakistanis whose lives would be shaped by the new boundaries, and the interplay between the political and the technical in the boundary-making process speak to a wider anxiety about the positionality of the geographer-as-actor and the geographer-as-witness, both in 1947 and in the present moment.

The story told here about Partition, geography and Oskar Spate is complicated by questions surrounding the imperial positionality of the geographer. I follow scholars such as Susan Bayly in thinking that during the middle decades of the 20th century – the era of late colonialism and decolonisation – the careers and writing of Western academics are interesting because they often 'challenge simplistic accounts of "colonial minds", "colonial intellectual projects", and colonial "civilising missions"', with the complicity of Western thought and scholarship in the travails of empire 'as a mere exercise in power-knowledge, conceived solely as a means of dominating voiceless Asian [colonised] Other' (Bayly, 2009, p. 192). Such accounts were integral to much anticolonial thought and rhetoric, of course, and have a similar relevance for postcolonial theory and scholarship. But at the same time, Bayly continues, such figures were still caught up in wider force fields of imperial power and cultural assumption and were not immune from their impress. In other words, anticolonial, and now post-colonial, thought begs questions about (rather than provides firm answers to) how academic disciplines like geography (or for Bayly anthropology) and those schooled in their ways were implicated in imperial power and resistance. Bayly has in mind the 1940s work of the French Orientalist scholar Paul Mus and his involvement in France's colonial crisis in Indochina. Oskar Spate's involvement in the Punjab Boundary Commission provides another interesting and overlooked case in point.

Geographers were sometimes vocal and articulate critics of empire and supporters of decolonisation. Spate at times was one such geographer. At the same time, he became caught up in boundary-making practices that were never, and perhaps could never be, fully disassociated from colonial discourses and imperial mind-sets about land, territory, borders and frontiers. In a very basic respect, the act of drawing an international border across, around, and between colonial populations and territories striving for independence, and as an integral part of the process of granting them sovereignty, was itself an imperial act. The geographical knowledge that facilitated such boundary-making, whether or not that knowledge was mobilised by geographers, politicians or lawyers, and to fashion one post-colonial outcome or another, was rooted in an epistemology that was forged in the image and long history of empire – of command and sovereign authority – and with nation-state territoriality as an alternate (if interrelated) model of sovereignty.

The chapter addresses two overarching themes – the modern-colonial rationality and pragmatism of the courtroom and what, according to Spate, and Congress and Muslim League actors, counted as a 'workable border', and the complicity of Western academics and their expert knowledge in anticolonial projects and processes of decolonisation. The first half of the chapter places Spate within the context of the Punjab Boundary Commission, outlining his prior work on the geographical questions posed by a potential future Pakistan, as well as his position as a consultant for the Muslim League delegation working in Lahore in July 1947. I then examine the work he was hired to undertake, which involved devising a claim for Qadian to go to Pakistan. The second half of the chapter explores some of the ways his geographical expertise was put to work beyond this remit, specifically to produce geographically sound rebuttals to competing claims from the Congress and Sikh delegations, introduced in the previous chapter. This work involved picking apart the methodology used to produce the Congress red map, as well as working on defence mapping and devising responses to the written claims of the Sikh delegation. I conclude the chapter with a discussion about the ways that all of this work was both incorporated and sidelined by the boundary commission itself, as well as the wider political and legal frameworks that shaped the boundary-making process more broadly.

Oskar Spate and Geography at the End of Empire

Oskar Spate (Figure 5.1) is recognised among geographers for his later foundational work on Southeast Asia and the Pacific, but when he was 36 years old, he travelled from London to Lahore in order to assist the Ahmadiyya community in preparing their case to the boundary commission. Spate was initially approached to lend assistance to the Ahmadiyya cause by Mirza Ali, the Imam of the Ahmadiyya community and Fazl Mosque in Wandsworth, London. As Spate

Figure 5.1 Professor O.H.K. Spate.
Source: Australian National University, 1975.

recalled: 'Could I help them to make sure that their sacred city, Qadian, stayed on the right side when the Punjab was partitioned?' (Spate, 1991, p. 47). The Ahmadis wanted a geographer to help them build their claim to Qadian. Mirza Ali sought his assistance upon the recommendation of Spate's well-connected LSE colleague Dudley Stamp: 'Obviously, this was Dudley Stamp's doing. He was quite an operator', Spate wrote (1991, p. 47). In his 1991 memoir, he noted how tenuous his expertise on the issue might actually have been, acknowledging that he 'knew nothing of the Ahmadiyya beyond the name, and had never heard of Qadian' (Spate, 1991, p. 48). He wrote that his 'one qualification' for assisting the Ahmadiyya community was that he 'had written what I believe to have been the first article on Pakistan by a professional geographer, sent from Bombay in 1943 on my monthly ration of nineteen airgraphs (a sort of microfiche) with maps on four air letters' (Spate, 1991, p. 48). However, Spate did have experience working and living in South Asia, having worked as a lecturer at the University of Rangoon in colonial Burma after he finished his PhD at Cambridge in the 1930s. He was in Rangoon when World War II began, and he was stationed there during the first Japanese air raid, in which he was badly injured and evacuated to India to recover. From 1941–1944, he worked as a military press censor in Bombay, but upon promotion to Major, he moved to the Inter-Service Topographical Department of Southeast Asian Command. By the end of the war, he was stationed in Kandy in colonial Ceylon.

Historians have tended to make good use of Spate's work to nuance their narration of the nature of the rushed and haphazard proceedings (Brotton, 2012; Chatterji, 2007; Chester, 2009; Jalal, 1994; Tan and Kudaisya, 2002). Geographers in the middle of the 20th century, such as S.P. Chatterjee, Ali Tayyeb (1966) and Nafis Ahmad (1968), and the more recent work of political geographers including Graham Chapman (2012) and Willem van Schendel (2005) have also been influenced by Spate's work on Partition. With William Gordon East and Charles Fisher, Spate produced a series of well-regarded edited volumes on the political geography of Asia titled *The Changing Map of Asia* (East et al., 1971). Spate's published articles have also provided a geographical authority on the geographical outcomes and implications of the Radcliffe Award. He argued that the final award favoured the Congress in the Punjab and the League in Bengal, and this assessment is generally accepted as an accurate geographical reading, and so he provides a certain amount of geographical authority for wider scholarly arguments about the politics of the process.

However, neither geographers nor historians have examined in detail Spate's papers, which are held in the National Library of Australia archives. Spate's technical skills as a mapmaker, his more specific geographical knowledge, and how he used both to influence and shape both the Muslim League claims and the wider debate during the partition process in the Punjab, have remained unexplored. One explanation for this is that Spate himself did not think of himself as central to the process, and after the award was published, he was disappointed and thought that his contribution had been minimal at best. He certainly may have downplayed his contributions or underestimated the power and privilege his position as a British academic geographer afforded him. Even so, the wider historiographical narratives of Partition have generally (with the notable exception of Chester (2009)) tended to focus on the negotiations and politics taking place beyond the courtrooms and boundary commission hearings. This focus on Spate therefore casts vital new light on the bearing that modern colonial categories and ways of representing territory spatially and cartographically had in the Lahore courtroom, not least through his influence on the debate.

Spate produced a small yet fascinating archive of his work relating to the Punjab Boundary Commission, including draft and sketch maps, statistical work, draft arguments in response to other parties' claims and a collection of letters. These materials shed new light on the geographies of the partitioning process itself, in addition to illustrating some of the contradictions inherent in applied and technical geographical work beyond the academy. Spate sought both 'objectivity' (an ideal which was informed by his role as an academic, as well as a concerned observer who was keenly aware of what was at stake) as well as a favourable outcome for his employers, the Ahmadiyya community and the Muslim League. While Spate himself worked to reconcile such tensions in his own mind, this chapter interprets Spate's materials through an engagement with those tensions. Similarly, Spate's work was filtered through the same colonial systems of knowledge and control within which the broader process of Partition was conducted, and some of the genealogies which I have outlined in Chapter 2 of the book. In this way, Spate's work was at times critical of empire, while at others, he was acquiescent to much of the imperially constructed discourse at work in 1947.

Spate's primary role in the boundary-making process was to provide a geographically sound justification for the Ahmaddiya's claim to Qadian, and for its inclusion within the future Pakistan. However, his geographical advice and expertise ran much wider and deeper than this. In part, this was because the Ahmadiyya delegation was closely aligned with the Muslim League delegation. In fact, the Ahmadiyya representative to the boundary commission was none other than Muhammad Zafrullah Khan, who also represented the League. Spate therefore ended up as an advisor to both the Ahmadis and the League. He produced a variety of sketch maps and written analyses of Congress Party, Muslim League and Sikh positions and claims, and commented at length on the various claims that went into the overall arguments of the Muslim League. He attended many of the hearing sessions in Lahore, recording his observations in his journal. After Partition, Spate authored two articles, one on the process of Partition, and the other on the future of Pakistan. Both were published just a few months after independence (Spate, 1947, 1948). Spate's foundational monograph in regional geography, *India and Pakistan: A General and Regional Geography* was first published in 1954, and updated extensively in 1967 with Andrew and Agnes Learmonth (Spate et al., 1972).

In some ways, Spate's archive is unsatisfactory. It does not clearly show us that geographical expertise or techniques could have or would have significantly altered the outcome. It certainly does show us that Spate himself felt quite powerless in the face of a complicated and large-scale geopolitical process and that his interventions were sometimes desperately small attempts to assist the League in achieving small positive outcomes. But as a geographical archive of Partition in the Punjab, Spate's materials show how geography as it was practiced in 1947 might have provided some solutions while ultimately remaining unable to overcome the fundamental issues of minority, representation, and territory that lay at the heart of the issue.

Spate on India Before 1947

Before 1947, Spate had never been 'west of Delhi and Bombay', but he was correct when he mused that his 1943 article on Pakistan was the most substantial academic treatment by a Westerner of the geography of Pakistan published at that time (Spate, 1991, p. 48). 'Geographical aspects of the Pakistan scheme' was written primarily for a British academic audience, who were generally unfamiliar with the idea of Pakistan. He examined two of the key geographical arguments made in favour of Pakistan. The first was an argument about the geographical bases of territorial delineation – an argument which grappled with issues of geographical determinism. Spate wrote that supporters of the Pakistan scheme argued that 'geographical factors combine with those of religion and custom to differentiate the Muslim areas from the rest of India so completely that none but an administrative unity, crushing all local interests, is possible' (Spate, 1943, p. 128). The argument in favour of Pakistan, which Spate said had 'a very strong bias towards geographical determinism', linked the Muslim majority in the north-west region of the subcontinent to a particular set of geographically conditioned cultural and religious practices which were distinct from the practices of Hindu communities further south, and worked to justify the notion that Muslims constituted a separate 'nation' in India (1943, p. 128). Some of these social and cultural factors included occupation, dress, custom and a historical association with and preference for the Urdu language, although the political mobilisation of these is a part of the long and complicated history of communalism and nationalism in South Asia.

Spate notes that population maps could be, and had been, mobilised by supporters of Pakistan to illustrate this territorial relationship; he cites Rahmat Ali's work (discussed in detail in Chapter 3) and 'El Hamza' (1942) as two writers who used population maps extensively in their arguments in favour of Pakistan. These 'population maps' were informed by and often based upon maps produced by the government, in the *Imperial Gazetteer* atlases for example, which depicted simple population majorities using categories captured by the census; the primary category deployed by proponents of Pakistan was, of course, religion, although the government produced maps depicting race, language and caste, among other categories. However, these population maps, and the more general geographical determinism argument for Pakistan, could not fully explain the large numbers of Indian Muslims who lived outside of what Spate termed 'Pakistan proper', and whose social and cultural practices did not entirely reflect those of the Muslims in the north-west (1943, p. 128).

Spate wrote that the second argument, which circulated among supporters of the Pakistan idea, appealed to historical and political geography. It was that, historically, India had never been unified under a single government or state. Spate argued that proponents of Pakistan were correct in that 'in the past India has rarely been united'. Yet he continued: 'The unstated inference that this fact supports a north-south division of the Indo-Gangetic plain however by no means follows' (1943, p. 129). Indeed, it was often the case that physical and historical boundaries in India tended to divide the east and west, rather than the north and south. Consequently, Spate reasoned that the 'true' foundation of the Pakistan claim was not actually geographical but communal. According to Spate, it was demographic and economic concerns that lay at the heart of the Pakistan demand. However, 'the arguments for Pakistan', put forward by supporters of the idea including the Muslim League, were 'to a large extent [still] avowedly based on geographical considerations', making the issue one of interest and significance to geographers (1943, p. 125). Using the 1931 census data, Spate sought to demonstrate that if Pakistan was to be created, the Punjab and Bengal would most likely be subject to what he called a 'rectification of frontiers', with some districts of the Muslim-majority provinces going to India (1943, p. 130). His reasoning was contrary to Jinnah's claim to Muslim-majority provinces in their entirety, and it was not just Spate and Jinnah who recognised that such differences in position and argument were rooted, in good measure, in the problematic nature of the 1931 census data. Kenneth Jones (1989) notes that from the late-19th century onwards, religious leaders grasped that the census was a powerful tool with which to claim and access political influence. Religious leaders gained and wielded this influence by campaigning to convince as many people as possible to respond to particular census questions in specific ways. By the 1940s, the Indian elite recognised that the census was a battleground of political persuasion and influence in this and other ways, and a significant portion of the Congress case rested on the unreliability of the census. In his 1943 article, Spate prefers the phrase 'rectification of frontiers' to the term 'partition', and so doing alludes to the failed 1905 attempt to partition Bengal for administrative and revenue purposes. He further notes that 'although "Partition" has a bad ring since Curzon's failure a division under purely Indian auspices might be a different story' (1943, p. 132).

Yet, for Spate, the problem with the conception of Pakistan was not over the delineation of boundaries or with the possibility of a geographical partition, but with population distribution: 'There remains the larger question of the extent to which the creation of Pakistan would in fact represent self-determination for the Muslims' (1943, p. 132). According to the data that Spate had before him, a third of British India's Muslim population would be left in India after the creation of Pakistan, leaving a 'scattered minority'. Alternatively, the new Muslim state would have a sizeable non-Muslim population of between 40% and 46%, 'minorities too large for successful assimilation' (1943, p. 132). The solution to this problem – what Spate meant by 'the rectification of frontiers' – was not easy. Without a partition of the Punjab and Bengal, the new Pakistan would have too many non-Muslims for a state designed to be a 'homeland for Muslims'. But with a partition, the new Pakistan would leave too many Muslims in India. Both of these solutions would fail to adequately fulfil the purpose of Pakistan as a homeland for India's Muslims. Either way, Spate said, Pakistan would be 'incomplete'. 'This mere fact of its incompleteness is in itself probably the strongest argument against the Pakistan idea', Spate surmised (1943, p. 132).

Spate's work on Pakistan anticipated many of the geographical features and problems that faced the boundary commission. He was already familiar with the issues surrounding the census data and aware of the communal issues at the heart of the Pakistan problem. He recognised to a certain extent the very real conflict between the political demands associated with the idea of Pakistan and the territorial and geographical realities of its implementation.

Spate and the Boundary Commission

Spate left London for India on Saturday, 12 July 1947, travelling by steamer via Sicily and Cairo. He worked in Lahore, out of his room at the luxurious Faletti's Hotel near the city's governmental hub, and occasionally at the home of Muhammad Zafrullah Khan, where the Muslim League's delegation was based. He worked long hours, often late into the night, and he took care in his production of draft maps, calculations and assessments of the competing claims. He enjoyed Lahore very much, writing that the city, was 'incomparably the greatest centre of Islamic culture in India', with 'bookshops which would put an English city of the same size to shame'. As the capital city of the Punjab, it was the 'most dynamic and self-confident Province of British India' (Spate, 1991, p. 52). He appreciated both the historical and cultural significance, as well as the political implications of the partitioning of the Punjab. It would be the division of 'the Land of the Five Rivers, guarding the gates of India, traversed by the great highway used by armies from before Alexander the Great to the Persian Nadir Shah in 1739, and the core both of Mogul power and that of the Sikhs, the last of the Country Powers to yield to the persistent encroachments of the British' (1991, p. 52).

Despite disagreeing with the overall decision to partition India, he believed the Muslim League's territorial claims were valid. He wrote in his journal, 'I do not like Pak(istan) in itself but realise its inevitability; and so it is important to make it as good a working unit as possible with good working frontiers. The moderation of the Muslim claim makes my personal position much easier as I am firmly convinced in my own mind, once given Pak, of the essential justice of their claim' (19 July 1947, Spate Papers). Later, he wrote: 'once given Pakistan (an important qualification), the Muslim case seemed to me entirely legitimate. There was thus never the slightest conflict between my duty to my employers and my sense of professional fitness' (Spate, 1947, p. 201). He also respected the leaders of the Ahmadiyya community and noted the importance of the Punjab for Pakistan. 'The Punjab is Pakistan's only riches, Congress has virtually the resources of all India', he wrote in his journal (26 July 1947, Spate Papers). He saw his role as both assisting the League in their own claims and simultaneously mitigating the wider potential problems associated with a poorly drawn boundary, which he believed were compatible goals.

Spate's articulation of this compatibility and its enabling of his support for the Muslim League's territorial claims illustrates the paradox he was working within in his role as a purported 'expert'. In his capacity as an academic, Spate was expected to remain at an emotional and political distance in order to maintain both objectivity and impartiality. 'Taking sides' and becoming too politically involved would render his technical expertise and advice to his employers less trustworthy. For example, what would he have done had the League's claims been geographically unsound, as he argued many of the Congress claims were? Yet his desire for objectivity was itself rooted in the problematic and suspect claim that academics could, or should, be expected to strive for complete neutrality. The wider context in which Spate had only recently been working as an academic-turned-military geographer (the Second World War) was a series of highly charged geopolitical circumstances. Lives were at stake. Spate was moved to participate in the boundary commission in some way and to do his best on account not only of his professional reputation but also his fair-mindedness.

In examining the boundary commission and Muslim League documentation, it is clear that Spate applied his particular geographical expertise to his contributions to the League's case. He valued rationality and evidence-based reasoning, and he often criticised the other claims in these terms. But this emphasis on rationality and evidence was not, in his mind, in contradiction with his hope that he might influence the final outcome and that this might be positive for the people involved. In particular, he was concerned about escalating violence. Unlike Thomas Holdich, for example, Spate was a more ambivalent participant in empire, critical of its violent tendencies, while also being complicit in it due to his background education and career. His education and experience were both what facilitated his critical perspective on the late stages of the British Empire and what afforded him the privilege and opportunity to take part in it. Interestingly, he was put in this position of expert again, advising the British Government over the future of Fiji and regarding his 1959 report *The Fijian People* – one of his most incisive pieces of work. Like other British, French and American geographers of his generation, Spate was bound by a late imperial age and mindset while being critical of aspects of it and was therefore sceptical of his own ability to have a direct impact on the boundary commission.

By the time Spate arrived in Lahore, the League and the Ahmadiyya representatives had already developed the majority of their arguments. Spate's task was to strengthen those political arguments by developing sound geographical justifications for them. He ended up providing more general assistance in adding points to the League and Ahmadiyya testimonies and in providing geographical rebuttals to the other arguments. He did this partly because he felt compelled to help for the reasons outlined above, and partly because he was also interested in working out some of the geographical and cartographic puzzles thrown up by the other parties' competing claims. Over the course of the week, he produced sketch maps, including strategic maps outlining issues of defence, a map showing the proposed boundary in relation to railways and possible corridors connecting Amritsar to India, and some general population maps. He also produced population (age-sex) pyramids to show that 'on pretty well every demographic index Muslims show as most stable element of poptn' in the Punjab (25 July 1947, Spate Papers). In one of his more interesting contributions, discussed in more detail below, Spate produced a detailed response to the Sikh case: 'over 7 pages and I think I can honestly say that they are what I would have written as a purely academic exercise, uninfluenced by the fact that I am retained by the other side', he wrote in his diary (20 July 1947, Spate Papers). He also provided the most robust geographical critique of the Congress red map (discussed in Chapter 4), and he used some backward calculations to determine the method their mapmakers had used when mapping their claim. He insisted that the original Muslim League claim was 'very modest', although 'the correct one on the whole'. Its political effectiveness, however, was a different matter: 'from a bargaining point of view their maximum isn't maximised enough!' (19 July 1947, personal diary, Spate Papers).

Spate's source material is not easy to deduce, although it is clear that he drew heavily on official census records from both 1941 and 1931 for his population maps and age-sex pyramids. A number of extracts from the census exist in his personal files as part of his notes. He also used the census records to test his hypotheses regarding the Congress method in constructing their red map. He left London in a rush, noting: 'Somehow on Saturday morning I got the visa and a couple of books on boundary-making from the Royal Geographical Society' (Spate, 1991, p. 49). He does not record which books these were but does mention examining 'a report on the Turkey-Iraq bdy. comms. of 1924' which he found 'dull and not very useful' (19 July 1947, personal diary, Spate Papers). He also mentions Vittorio Adami's *National Frontiers in Relation to International Law,* published in 1927, with regard to the importance of visible boundaries.

Spate notes that he had already begun work on his 'big geography' of India, which suggests that he arrived in Lahore with some material, as well as previously conducted research. He also probably brought with him the maps that he had used for his 1943 article; and upon arrival, he worked with a group of geographers from the geography department at the University of the Punjab at Lahore, headed by Dr Kazi S. Ahmad. That team drafted the final maps which were officially submitted to the boundary commission. In a letter, Ahmad originally gave Spate permission to name him in his publications and lectures, but later requested that Spate refer to the department without naming individuals, which is perhaps one of the reasons why the role of the university's geographers in the negotiations has been obscured and is sidelined in historical accounts (Cheema, 2000).

Throughout the boundary dispute, the relationship between the political debate and the geographical debate was evident in the cracks in and between different Partition plans. Stuart Elden (2013) reminds us that 'territory' is constructed through the negotiation of the political and the technical, and the refashioning of colonial territory in the subcontinent into India and Pakistan was no different. Spate wrote that the Muslim League claim 'was a good deal less than my notional division in the 1943 article' (Spate, 1991, p. 54). According to Spate, the League argument, shown in Figure 5.1, 'stuck on the whole to population; they claimed the Bist Doab and a riverain strip on the left bank of the Sutlej' (Spate, 1991, p. 54). This claim, according to Spate, was sound according to the geographical principles he recognised. Their claim to both Lahore and Amritsar was made on the grounds of contiguity. Similarly, the claim to Qadian (of special interest to the Ahmadis and analysed in detail in the next section) and Gurdaspur, a district in the north on the Kashmir border with only a scant Muslim majority, was made on the basis of contiguity as well as a claim to infrastructural unity because it was the site of a major canal headworks. When Spate arrived in Lahore in the final days before the hearings began, he knew and noted that 'the Boundary Commission had already started preliminary work', and believed that the official League claim was close to the ideal technical boundary (Spate, 1991, p. 48).

The Congress-Sikh claims, on the other hand, did not match the geographical borders which Spate (nor Radcliffe, it turned out) recognised in terms of territorial and political logic. These claims were 'amazing', 'and 'supported by an amazing map', 'a giant gerrymander'. Here, he was referring to the red map (Figure 4.3), but he was sceptical of the claims as a whole. Spate saw them as 'extreme', and indicated that it was 'an ambit claim by Congress (one cannot be too sure for the Sikhs)', an attempt to gain the best outcome for India (note on possible corridors, Spate Papers). The disputed area, between the blue (Muslim League) line and the red (Congress-Sikh) line, is illustrated clearly on Spate's draft map of the Punjab (Figure 5.2).

Spate was not convinced by either the Congress or the Sikh political arguments, and his attention was more focused on demographics, infrastructure (railways and canals) and defence, which he identified as the most important geopolitical issues at stake after the population question. He did, however, recognise the potential for the Congress and Sikh claims to skew the final result in favour of India should the boundary commission pursue a political tit-for-tat approach (which Radcliffe did, in the end, adopt), and worried about the potential political impact of Congress strategy on the geographical concerns of the dispute. The Congress-Sikh claims, he said, would 'cut Pakistan's throat' (19 July 1947, personal diary, Spate Papers).

For this reason, and despite the fact that he was hired primarily to work on the claim to Qadian, he was determined to provide a strongly geographical response to Congress and Sikh arguments for Zafrullah Khan to use in his oral testimony. He found himself analysing the geographical components of a debate which was, on all sides, a legal battle. The Congress and Sikh delegations did not have a Western academic geographer working on either of their cases and so Spate's arguments, while prepared for the Muslim League and based on geographical practice, were not part of a geographical battle. Rather, his geographical expertise was inserted into a bigger political and ideological battle which extended far beyond the realms and remit of academic geography. His work was, therefore, tangential in a sense, simply because the geographical discourse that framed his drawing of potential boundaries and assessing boundary claims was marginalised by the legal discourse which underpinned the boundary commission process.

This legal discourse was informed in large part by the *process* of the boundary commissions, which were conducted mainly in courts of law, and overseen by

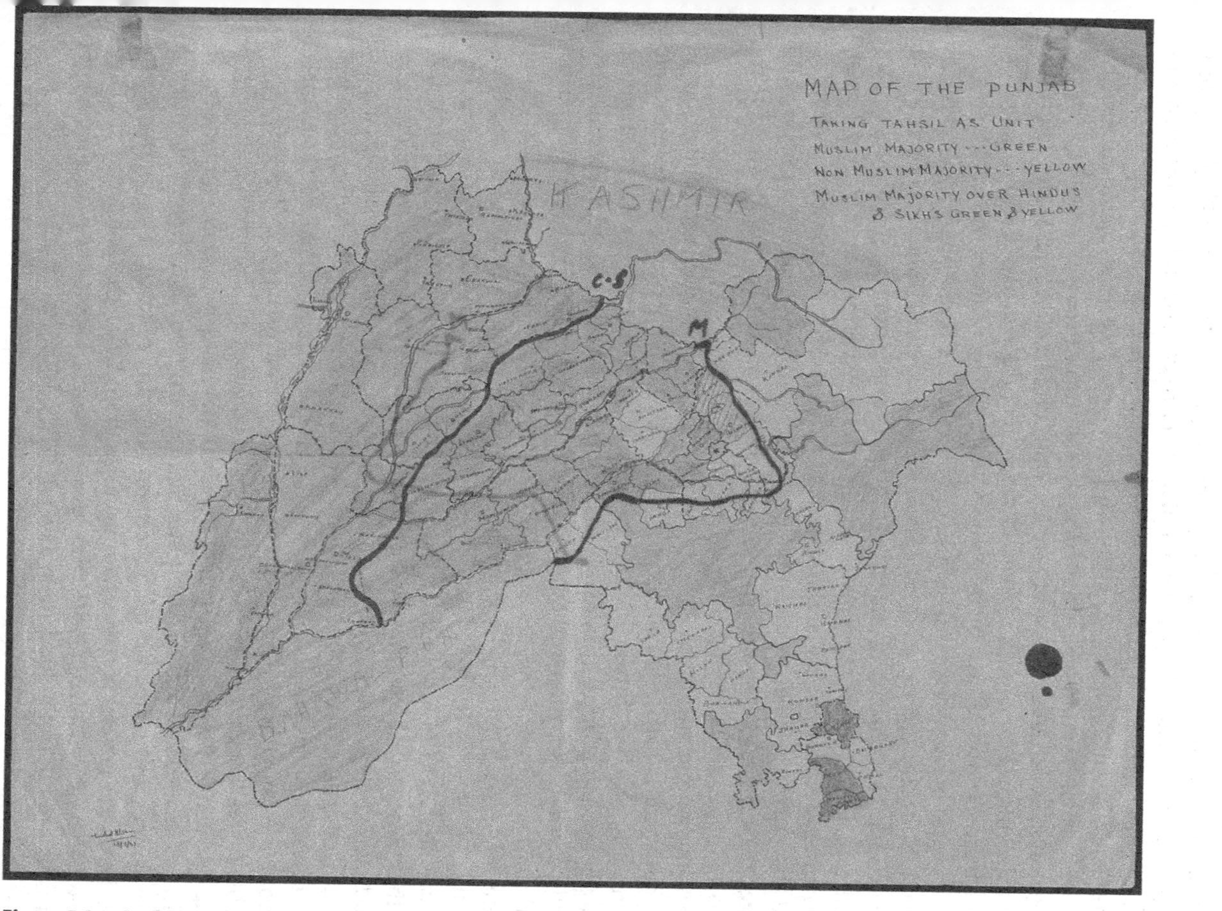

Figure 5.2 Draft Map of the Punjab from Spate's Collection. Depicting Muslim and non-Muslim majorities by *tehsil*, with Spate's addition of the Congress-Sikh and Muslim League proposed boundaries.

Source: The Papers of Oskar Spate, MS 7886, National Library of Australia.

judges, and was in many respects dictated by the British government's emphasis on the *legality* of the transfer of power. Indeed, when Spate wrote that the 'Ahmadiyya were the only people concerned with the Partition to whom it occurred that geography might have something to do with drawing boundaries', he was referring in large part to the British government (for which he reserved most of his criticism regarding the process) (Spate, 1991, p. 57). In a press conference on 5 June 1947, for example, Mountbatten said, 'In the process of quitting power in India, we must try and approach it in as legally correct a manner as possible' (Sadullah, 1983, p. 131). The government held a referendum in the provinces which would be partitioned to determine if there was 'popular' support for Partition. The delegation testimonies were often concerned with arguments over the interpretation of the terms of reference, the legality of utilising census data which was well-known by those involved to be inaccurate, the legal rights of property and business owners, the use of authoritative texts (often legal texts, for example, David Lloyd George's *The Truth About the Peace Treaties* and a volume of *International Conciliation*, produced by the American Branch of the Association for International Conciliation in 1919), and the general insistence (by the British government) on *compromise* and *negotiation*. Additionally, it is not surprising that legal discourse prevailed because of the prominent role played by lawyers and judges not only in the boundary commissions but also in the longer struggle for Indian independence more generally.

But this dominance of legal discourse also speaks to the broader philosophical and historical construction of territory through the application of political and technical knowledge. Stuart Elden has written extensively about the relationship between law and geography in the construction of territory as a politically and technically derived 'extension of the state' (Elden, 2013, p. 322). He notes that, particularly in the colonies, the development of geographical ideas of measuring and administering territory 'led, or were partnered by, developments in legal and technical practices' (Elden, 2013, p. 326). The boundary commission hearings indicate that, by 1947, both the nationalist elite and the colonial government had, for the most part, agreed upon a cartographic and administrative mode of understanding and governing the territory they all knew to be 'India'; what was at stake was a legal struggle over competing narratives of nationhood, all of which were built on this same construction of 'Indian territory'. Indeed, Pakistani nationalist discourse often struggled to assert a coherent claim to territory because it was unable to challenge the authority of the perceived geographical unity of 'India', creating the 'contradictory' discourse of Pakistan that historians so often highlight.

In fact, the leaders of the League and Congress agreed that, after independence, all provinces and states would be required to join either the new Indian or the new Pakistani government, as Mountbatten noted, 'for the good and sufficient reason that they did not wish this plan to encourage what I might call the Balkanization of India' (Sadullah, 1983, p. 132). Importantly, this policy encompassed the princely states, of which there were over five hundred in the subcontinent, and whose rulers maintained autonomy from the British through a system of treaties. Independence for India and Pakistan also meant the dismantling of the political and economic relationships between British India and the princely states for the purpose of consolidating British and princely India's provinces under the two new territorial nation-states (Copland, 1991). The physical, economic, political and social value of the territory of 'India' was generally assumed, while the question of sovereignty was often rooted in debates about the legality of historical and cultural claims to that territory. For example, the Congress representative noted during his testimony, 'The river waters are from the economic point of view, an important asset of the province as a whole, to the building up of which the whole province has contributed. Now, the province is being divided. Is it not fair that a part of this asset should be made available to the Eastern province for its economic security?' (Sadullah, 1983, p. 37). Additionally, he called upon a historical account of the relationship between the Sikhs and the Punjab, saying, 'They are really the builders of the Punjab'. He said, 'It appears that at one time they conquered not only what is now the whole of the Punjab but even a part of India what is now known as NWFP and other areas and it is from them that the British in 1840 got the whole of the modern Punjab' (Sadullah, 1983, p. 47).

The introduction of 'other factors' in the Commission's terms of reference considerably broadened the scope of the debate, which did not leave any historical, geographical, religious, social or economic aspect of the Punjab untouched. The proceedings thus present a fascinating study of the Punjab, its people, land, mountains and rivers, the implications of its rail and road communications, the network of canals, electric power transmission, ownership and colonisation of lands and the development of its trade and industry (Sadullah, 1983, p. iii).

'India', and 'the Punjab', as geographical and territorial entities capable of being administered, studied and partitioned, were not in dispute during the process. Nor were the identity categories that the government and the political parties applied to the Indian population in dispute (although both of these forms of knowledge are critically scrutinised and challenged by academics today). The scope and potential, therefore, for Spate's technical knowledge to produce a radically new vision for ordering territory through the partitioning of the subcontinent, and by extension what role geographical discourse and analysis played in the boundary-drawing process and subsequently in Partition historiography, needs to be placed in this context. In spite of his insightful criticisms of the boundary-making process, and of imperialism's penchant to reduce the question of territory to the cartography and interests of the state, Spate's technical geographical knowledge was rooted in this same logic and lineage of territory-making.

In terms of 'imagining' a new India, and perhaps more significantly of envisioning a new Pakistan, both the British government and the nationalist leaders of the Congress Party and Muslim League were working from a historically constructed blueprint of imperial territorial sovereignty that had been scientifically imagined, mapped and legally codified from the 18th century. Spate's criticism and scepticism was caught in a discursive loop. If the partition process was not, in fact, a technical geographical process of state-making, but was rather a political negotiation over nationhood, then the partition itself was a technical geographic fix for a political problem that was construed in particular ways. The political problem at hand was how to accommodate competing economic, cultural and religious claims to territory as it was imagined, produced and governed as components of empires and states, and how boundaries could then be demarcated and realised by the cartographer and surveyor. For Spate, however, the political problem to be adduced in the boundary commission proceedings, and settled by the technology of the map and survey, was wider and more fluid. It was about the relationship between this state-centric way of drawing lines across territory, and wider and competing conceptions of particular territories as real and imagined, archaic and contemporary homelands, and of competing histories of dwelling, displacement and geographical struggle. Indeed, Spate pointed to this epistemic tension, and how it reverberated a few months later in a lecture presenting an overview of Partition at the Royal Geographical Society in London, observing:

> The terms of reference were hopelessly vague: "To demarcate the boundaries of the two parts of the Punjab, on the basis of ascertaining the contiguous majority areas of Muslims and non-Muslims. In doing so, it will also take into account other factors". The inaccurate use of the word "demarcate" is symptomatic of the general vagueness; no one seriously envisaged the learned judges running around the Punjab with theodolites and concrete markers; but as the term was accepted on all hands I could only suffer in silence each time it was used, which was very often. (Spate, 1947, p. 205)

Spate became embroiled in a discursively asymmetrical debate. His technical evidence was deployed by the League in ways that were politically necessary but that fell short of their potential to open up what it meant to 'demarcate' to a discussion and analysis that did not just revolve around the legal passage from imperial to nation-state territoriality. Spate recalled that he 'suffered in silence' in the knowledge that there were less vague ways of demarcating boundaries, of overcoming the general vagueness that sprang, in his mind, from the courtroom separation of the map (and the violence of cartographic abstraction) from more complex and fluid conceptions of territory. The vague terms of reference that Spate beheld in Lahore lent themselves to political manipulation and manoeuvring, and with the kind of geographical expertise and imagination he could offer, and which could potentially incorporate many factors, subordinated to state calculations of territory. However, in Lahore, in 1947, Spate fully grasped the need to think and analyse strategically within the vague limits, or political limits, of the moment.

The Case for Qadian to Go to Pakistan

While the Congress and Sikh cases had to rely on the 'other factors' clause, the Muslim League's main argument rested on the population question. The Muslim majority areas in the contested section of the Punjab tended to correspond to the areas that Spate argued should go to Pakistan for strategic reasons, and so this particular aspect of the Muslim claim was not difficult for him to justify using the evidence he had to hand. More difficult for him were the two most highly contested areas. The first was the area he had initially been hired to produce a claim for: Gurdaspur district, where Qadian was located and which had a slim non-Muslim majority concentrated in the western *tehsils*. The other was Amritsar district, which sat at the heart of the Sikh religion and was central to the League's claim, despite the fact that the non-Muslim majority in Amritsar was not contiguous, but rather constituted a non-Muslim 'pocket' in a Muslim-majority area. A few days into the hearings, Spate wrote in his journal, 'Beginning to be nervy about Gurdaspur, think they may get at least Amritsar wedge' (23 July 1947, Spate Papers). He had, of course, been hired specifically to help the Ahmadis argue for Qadian's inclusion in Pakistan; they knew it would be difficult and Spate fretted over the fate of Gurdaspur, writing in his memoir about the final days before the award was announced, 'most of my time was spent in thinking up fancy precedents for extra-territoriality, enclaves, condominia, free zones, corridors and what have you, all to secure Amritsar and Gurdaspur Districts to Pakistan while ensuring Sikh access to their Golden Temple in Amritsar' (Spate, 1991, p. 57). I will return to these 'precedents' and Spate's work on enclaves and corridors below.

The issue of majority and minority populations is particularly relevant in a geographical analysis of the partition process. In chapter 3, I drew on recent scholarship to show how the issue of minority did not simply appear in 1947, but was an issue for Indian Muslim thinkers from the 19th century. And in the previous chapter, I discussed the ways in which the cartographic representation of population constructed a politico-spatial relationship between majority groups and territory in India, effectively erasing minority groups from the map and obscuring the more complex intercultural and interreligious social relationships which existed on the ground. The historically constructed cartographic claim to Pakistan put forward by proponents of the Pakistan idea emphasised that politico-spatial relationship in forming their arguments. It was this imaginary of populations as homogeneous and discrete social units, each of which could (and should) be tied to a corollary territorial unit, that underpinned the partition process. However, this issue was dealt with asymmetrically in the official arguments. The Muslim League relied on the authority and legitimacy of the 'majority population' to dictate its territorial claim, while the Congress and Sikh parties attempted to undermine this authority by stressing the relevance of 'other factors'. The question of population had been greatly significant in other partitions, including some of the most well-known partitions (in Ireland and Palestine, for example), begging the question: Was population as effective and equitable a tool as geographers and politicians believed it to be, at least for the purposes of geographical partition?

We saw in Chapter 2 that American geographers involved in the boundary-making processes in Europe after the First World War were deeply concerned with ethnic mapping as a means of addressing population issues. Partition in Ireland in 1920 was conducted partly on the basis of religious majorities and minorities, while the United Nations Partition Plan for Palestine dealt specifically with the issue of population in its attempt to fashion the new Jewish state. The issue of population was not mobilised in these different partition projects and processes in the same way (although Faisal Devji (2013) draws connections between Europe's Jewish minorities and India's Muslim minority in *Muslim Zion*). Rather, the quantitative and qualitative aspects of 'the population issue' were filtered through the specificities and complexities of the historical, geographical and cultural contexts within which geopolitical alignments and arguments were formed in different places and cases. However, population was considered a useful category to mobilise in varied partition contexts, and particularly so during the reorganisation of territory along nationalist lines during decolonisation processes (although there are exceptions). Additionally, the relationship between population and partition as a facet of decolonisation was quite a new phenomenon in 1947 and one with close affinities to longer and wider arcs of anticolonial struggle, and thus, as Spate intimated at the RGS, a relationship that was open to debate and contestation rather than a necessarily clear, fixed or authoritative reference point or precedent for adjudicating what Marxist-Leninist tradition and anticolonial movements termed 'the national question'.

As Spate sought to relate in the sketch map of Qadian (Figure 5.3) in his notes, the Ahmaddiya claim to Qadian would be difficult to defend because while Qadian had a Muslim majority, it was located in the contested Gurdaspur district, on the boundary line between Batala and Gurdaspur *tehsils*. Qadian is marked on Spate's map as 'K'. The town was established by the Mughals in the 16th century, and was the birthplace of the founder of the Ahmadiyya community, Mirza Ghulam Ahmad, and the community officially made Qadian the seat of its authority and the home of the Ahmadiyya leadership in 1908.

Qadian was 'a Muslim enclave' located about 55 kilometres north-east from Lahore and Spate observed that the place was 'a petty state within a State – crimes were reported to the Ahmadis before the police' (Spate, 1991, p. 55). Like Amritsar for the Sikhs, Qadian was the political, spiritual and historical hub for the Ahmadis, and the community was deeply concerned over the possibility that it might not be awarded to Pakistan. After the boundary commission hearings finished, Spate travelled to Qadian, where he delivered a lecture on political geography and met with leaders and members of the Ahmadiyya community. He found Qadian 'fascinating...right out of the Old Testament'

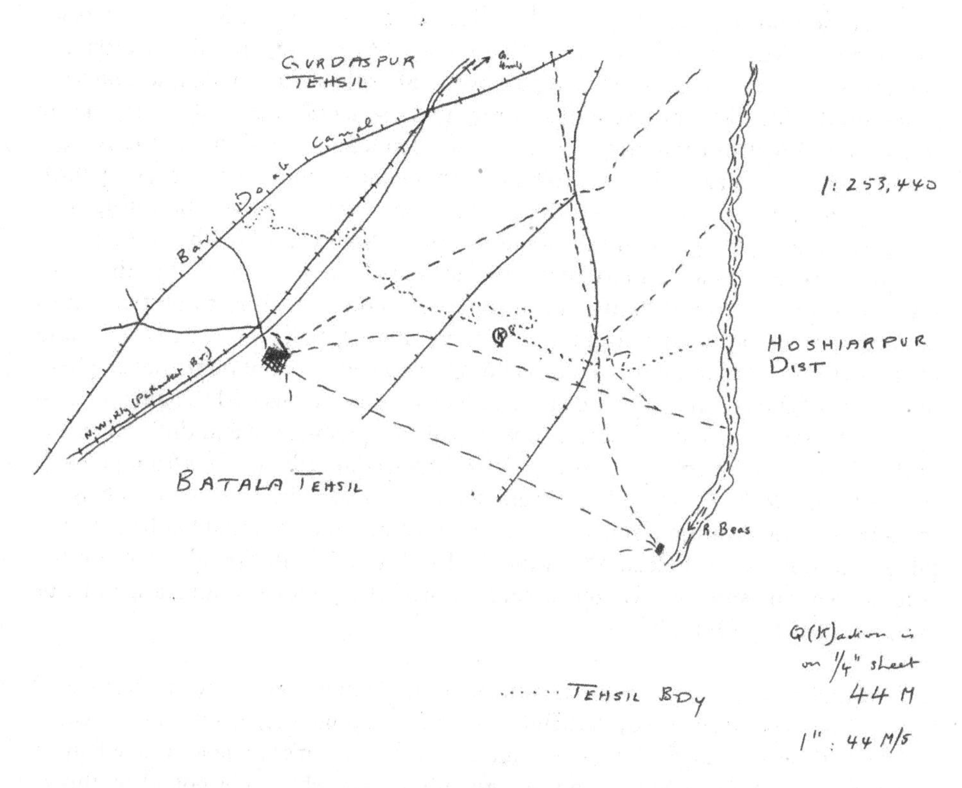

Figure 5.3 Spate's Sketch Map. Showing Qadian (K), close to the boundary (depicted by the small dotted line) between Batala *tehsil* and Gurdaspur *tehsil*. The map also depicts the River Beas along the eastern boundary between Gurdaspur District and Hoshiarpur District, and railway and canal lines.

Source: The Papers of Oskar Spate, MS 7886, National Library of Australia.

(Spate, 1991, p. 55). Of the Ahmadis he wrote, 'the community is very vigorous... there were many guests, Ahmadi scholars and pilgrims from all over the Islamic world' (Spate, 1991, p. 56).

The Ahmadis presented their case, specifically regarding Qadian, to the boundary commission on Friday 25 and Saturday 26 July 1947. They made what they called 'special claims' to Qadian, although they did argue that Qadian fell within the contiguous Muslim-majority *tehsil* of Batala. The Muslim League claim included Batala *tehsil*, as well as an adjoining non-Muslim-majority *tehsil*, Pathankot, where the canal headworks, which served the Ravi River and ran mostly through West Punjab, were located. The Congress had argued that Qadian was a Muslim pocket in an otherwise non-Muslim-majority Batala *tehsil*, but the Ahmadi representative, Sheikh Bashir Ahmad, responded to the Congress argument with what became a common argument: that 'you should employ a uniform method before you proceed to decide whether it is a Muslim-majority area or a non-Muslim majority area' (Sadullah, 1983, p. 242). In other words, the Congress claim to the contested Gurdaspur *tehsils* of Batala, Ajnala and Narowal rested on the fact that the Muslim majority in these *tehsils* was really just a highly concentrated Muslim-majority in a few discrete locations within the *tehsils*, surrounded by an otherwise Hindu-Sikh area (that claim would, of course, only have had any validity had they applied that method consistently across the Punjab).

The Ahmadiyya argument also focused on the Sikh claim to what they called 'certain shrines' (Spate noted that the Sikhs had, in fact, listed hundreds of shrines in their memorandum), and if the Sikhs' concerns were to be taken into account by the boundary commission, then the Ahmadis' concerns regarding Qadian also needed to be treated as valid. Qadian was a site of pilgrimage, where 'members of the community flock to hear learned discourses and to gain spiritual uplift'. Qadian was 'a living centre to which all people, who live in different parts of the globe look for religious and spiritual knowledge' (Sadullah, 1983, p. 245). Furthermore, 74% of Ahmadiyya branches were located in what would likely go to Pakistan; 'the economic stability of the community requires that Qadian should fall within the area of Pakistan' (Sadullah, 1983, p. 250). In Spate's notes on the Ahmadiyya community's written case, he noted the political importance of this argument for the Ahmadis:

> A neutral arbitrator – which the Chairman of the Commission will be in effect – is not likely to be convinced by declarations of the paramount sanctity of a holy place. This may be absolute truth for one community; but the same will hold for the holy places of other communities, and thus to an outsider, who is not bound by this essentially subjective view, the two cancel each other out. . . Therefore if the arguments cancel out it is a fundamental plank of the Sikh platform which is lost, but only an ancillary one of the Ahmadiyya. (Brief note on Ahmaddiya Memorandum [sic], Spate Papers)

It was on the basis of details like this that Spate subsequently mused in a more general way at the RGS about different and competing conceptions of territoriality as inherently 'subjective' and not amenable to 'demarcation' in any straightforward way. In the end, Qadian went to India, and after Partition, when the Punjab was overcome by riots and unrest, communication and transportation infrastructure were damaged and circumstances were deemed too insecure by the Ahmadi leadership to retain Qadian as its hub, the community moved its headquarters across the border to Pakistan. But the story of the Ahmadiyya community does not end in 1947, and the colonial context that created the conditions for a territorial partition on the basis of the sociological category of the Muslim also created the conditions for the Ahmadiyya community to be rendered a religious minority in independent Pakistan (Saeed, 2018). Asad Ahmed argues that, in Pakistan, Ahmadi identity is 'paradoxical', in that it is both Muslim and non-Muslim at the same time. This is because the Pakistani state considers the Ahmadiyya community to be a 'non-Muslim minority', while the community itself considers itself to be Muslim (Ahmed, 2010). As the Pakistani state has since independence wrestled with the ways it defines the boundaries of Pakistani Muslim identity, Ahmadis have found themselves at the coal face of legal debates about their place within Pakistan (Kamran, 2016). Partition produced the conditions for the community, which originated in the subcontinent, to be excluded on both sides of the border, and, therefore, at 'home' in neither new state.

Amritsar: 'The Key to the Whole Boundary Problem'

Amritsar presented a problem for the Sikhs that was similar to that of Qadian for the Ahmadis, although Radcliffe awarded Amritsar district to India, drawing the border at Wagah, between Lahore and Amritsar. More importantly, Amritsar's economic and geographical position in the Punjab meant that it was extremely valuable. Both sides insisted the district should go to them. Spate argued, both before and after Partition, that Amritsar should have gone to Pakistan, both for reasons of contiguity and due to economic factors. Indeed, he wrote in his notes, 'Amritsar is the key to the whole boundary problem' (Brief note on Ahmaddiya Memorandum [sic], Spate Papers).

In fact, one of Spate's most fascinating sets of notes in his files relates to these musings on enclaves and corridors. The idea was never seriously considered by the boundary commission (and was neither mentioned in the hearings nor in the memoranda to the boundary commission), but Spate spent a great deal of time and energy trying to devise a way to make a corridor work. While the idea seems especially absurd now, there was no guarantee (in 1947) that the boundary commission would award an entirely contiguous region, either to India or Pakistan; it was possible that enclaves would be given to one side or the other, especially in a context like that of Amritsar city. As Spate wrote in his notes at the time, 'If A[mritsar] & T.T. [Tarn Taran] are to be included in E. P[unjab] as non M[uslim] maj[ority] tahsils, it is only just that the M[uslim] maj[ority] tahsils E[ast] of B[eas]-S[utlej] should go to W. Pj'. Amritsar district had a clear Muslim majority and was well within a contiguous Muslim area. Amritsar city, however, had a Sikh majority, and was the spiritual and administrative heart of the Sikh community. 'Should Amritsar and Tarn Taran tehsils be allotted to E. Punjab, the question of their connection with its main territory by a corridor might arise' (Note on possible corridors, Spate Papers).

Spate was not the only one to float the idea of territorial workarounds to rigid contiguity. Other alternatives like the corridors were offered at various stages. Ayesha Jalal notes that the Governor of Bengal suggested to Mountbatten that Calcutta, due to its economic significance to the province as a whole, become a 'free city', administered jointly by elected Hindu and Muslim representatives

(Jalal, 1994, pp. 266–267). This idea was rejected because of the precedent it might set for Sikh claims to Lahore in the Punjab. There had also been a proposal for a united independent Bengal, put forward in 1947 by the Bengali politician Sarat Chandra Bose. His proposal echoed in some respects Choudhary Rahmat Ali's cartographic representation of Bengal, unified on the basis of language. He also echoed the emphasis on a decentralised union of states that characterised Iqbal's vision, but he rejected religion or community as the basis for forming those states. Rather, socialism was necessary for overcoming the divisions wrought by politics conducted on the basis of communalism: 'I have always held the view that India must be a Union of autonomous Socialist Republics and I believe that if the different provinces are redistributed on a linguistic basis and what are called provinces are converted into autonomous Socialist Republics, those Socialist Republics will gladly co-operate with one another in forming an India Union' (1947, in McDermott et al., 2014, p. 544). He argued presciently that Partition 'will dissolve the existing linguistic bonds and instead of resolving communal differences will accentuate and aggravate them' (2014, p. 544).

On a fundamental level, suggestions like free states and corridors did two things. First, they challenged the notion of a border as the territorial marker of absolute sovereignty. Corridors would work by creating a strip of territory that, rather than restricting access entirely, would channel access to specific, highly contested spaces for those whose religious and cultural lives would be severely diminished. Second, these ideas provided an alternative to rigid conceptions of territorial contiguity in favour of what was considered a more accurate and precise division of population; this, of course, speaks to the conundrum at the heart of Partition more generally. When territorial contiguity conflicted with the possibility of ideal homogenised populations (and homogeneous on the basis of communal or religious categories, as opposed to other salient categories that cut across community), what was the more important component of the nation: *the territory* or *the population*? Interestingly, the notion conjures up Jinnah's proposal for a 'land bridge' which might connect East and West Pakistan; Jinnah had argued that territorial contiguity was vital for the survival of Pakistan. Jinnah's corridor was never seriously considered, most likely because the discourse of minority and majority populations had become the heart of the Partition plan, and had ruled out the possibility of an alternative organisation of territory which did not depend so heavily on population statistics. Although Spate did not believe that communal claims to territory would create a workable boundary when they contradicted strategic and population-based claims (as, he believed, the Congress-Sikh claim to Amritsar city did), he recognised that the boundary commission would likely give Amritsar city to India based on its non-Muslim majority, and as part of a tit-for-tat compromise (where Lahore would go to Pakistan). He also recognised that enclaves might be a geographical solution to the intractable contradictions inherent in the terms of reference, which stressed both majority population and contiguity.

Spate outlined three possibilities for such corridors, none of which were ideal (although in the context of the wider process, every available option was objectionable in at least one way): '(a) SW of Beas-Sutlej confluence (b) via Jullundur (c) down from Kangra' (Note on possible corridors, Spate Papers). These are shown on Spate's draft map, (Figure 5.4) and labelled *a*, *b* and *c*. The Jullundur corridor followed a railway, and the Kangra corridor followed the west bank of the river Beas. Neither was a particularly workable boundary, according to Spate; they made access to the Bist doab (the strip of land to the west of the corridors) nearly impossible for West Punjab and Indian access to the corridors would be very 'roundabout'. The third corridor, from the south-west of the Beas-Sutlej confluence, was the least objectionable of the three; it would require the building of a bridge across the Sutlej, but rail connections were generally simple. This corridor, however, cut across two 'very important' railways, the impact of which depended almost entirely on the rest of the boundary. Ultimately, as with Jinnah's East-West corridor, no enclaves were given to either side, rendering the need for a corridor obsolete.

In two papers published after Partition Spate argued that most of the geographical logic of the debate, logically equipped for dealing with the fluidity and complexity of the territorial issues, got lost amidst the political claims and manoeuvres, a sentiment which historians and other commentators have since echoed. Spate believed that the 'other factors' clause in the terms of reference was dealt with carelessly and irresponsibly (Spate, 1991). The 'other factors' at play became political bargaining chips rather than geographical considerations to be taken into account either when the population question was ambiguous or 'other factors' became critical to perception of the future survival of India and Pakistan. About the canals, for instance, Spate wrote:

> The central Punjab, where lies most of the canal development, is such a unit that any division of it cannot fail to inflict serious economic damage; and, if the Punjab is to be divided, broad geographical factors would suggest a division on or east of the Beas-Sutlej line. In fact, however, the disputed area extended from the Chenab to east of the Sutlej and included nearly half the population of the Province. From the start, therefore, what would seem to be geographically a rational division was ruled out by political considerations. (Spate, 1947, p. 202)

It was clear to Spate that the Punjab should be divided along the Beas-Sutlej line, which is roughly where the League claim placed it. This would maintain most of the integrity of the canal system while causing the least disruption to the railway lines.

The canal and railway systems had been a key element of the Punjab's development since 1850 and had been part of the British government's project to modernise and exploit the Punjab for its agricultural and military potential. The canals were developed as part of a plan to expand the agricultural output of the

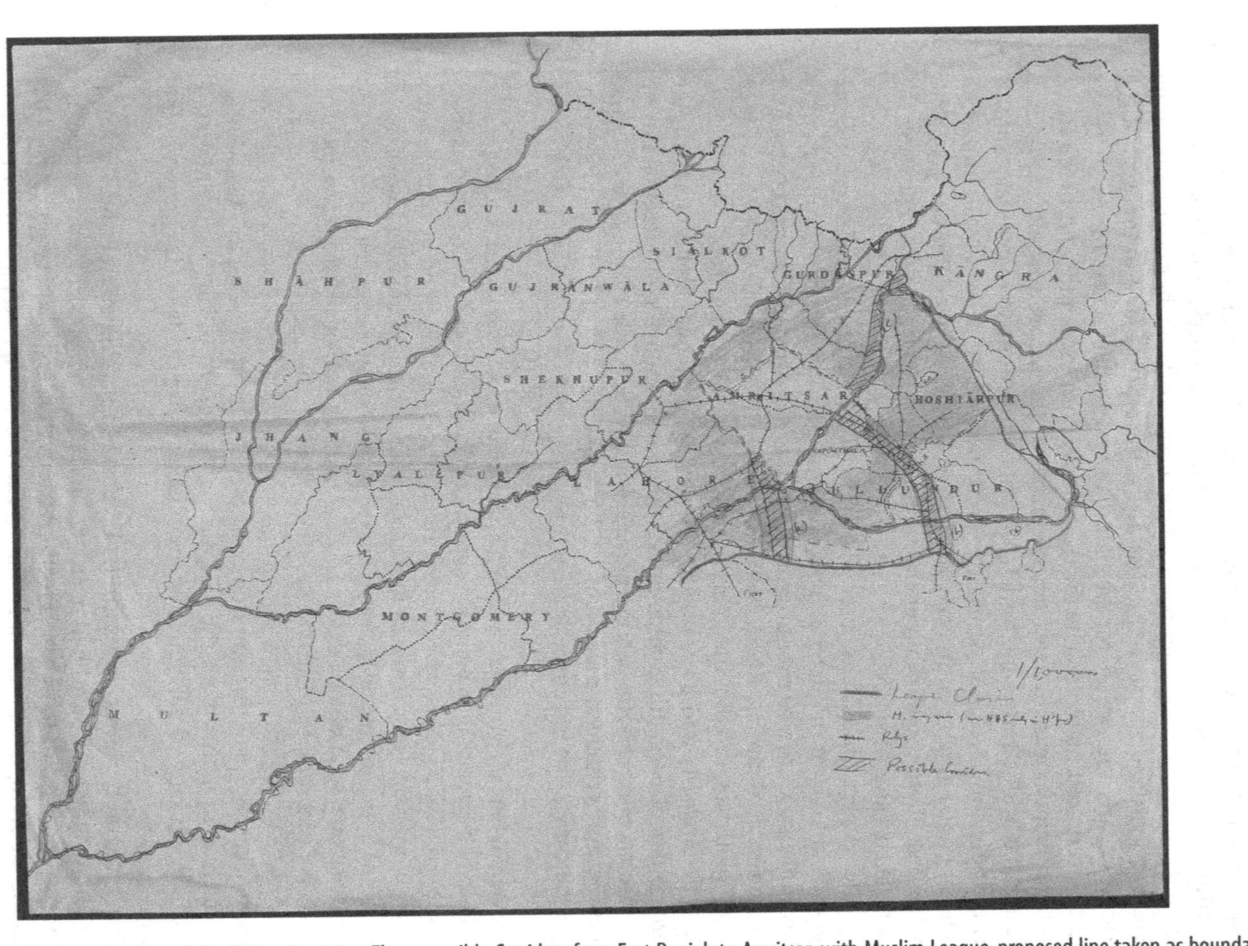

Figure 5.4 Spate's Draft Map Depicting Three possible Corridors from East Punjab to Amritsar, with Muslim League-proposed line taken as boundary and showing Muslim-majority *tehsils* shaded.

Source: The Papers of Oskar Spate, MS 7886, National Library of Australia.

region, and labour colonies, which consisted of village communities, were built in the western areas of the Punjab where migrant labour from the east could live. Alongside the development of the canals was investment in a railway network (Figure 5.5) which could connect the Punjab to the rest of Northern India, facilitating an expansion in the transportation of food, goods and labour. The Punjab was also a key component of the empire's defence strategy, a militarised buffer state separating the capital, Delhi, from the frontier zones further to the north. Both the canals and the railways had been designed specifically to unify and connect the Punjab, both internally and externally, to the rest of British India. These relatively well-connected systems (at least in comparison to comparable infrastructure in other parts of British India), depicted on this map were geared to unify, not divide, the province and cross religious and ethnic groups and sects.

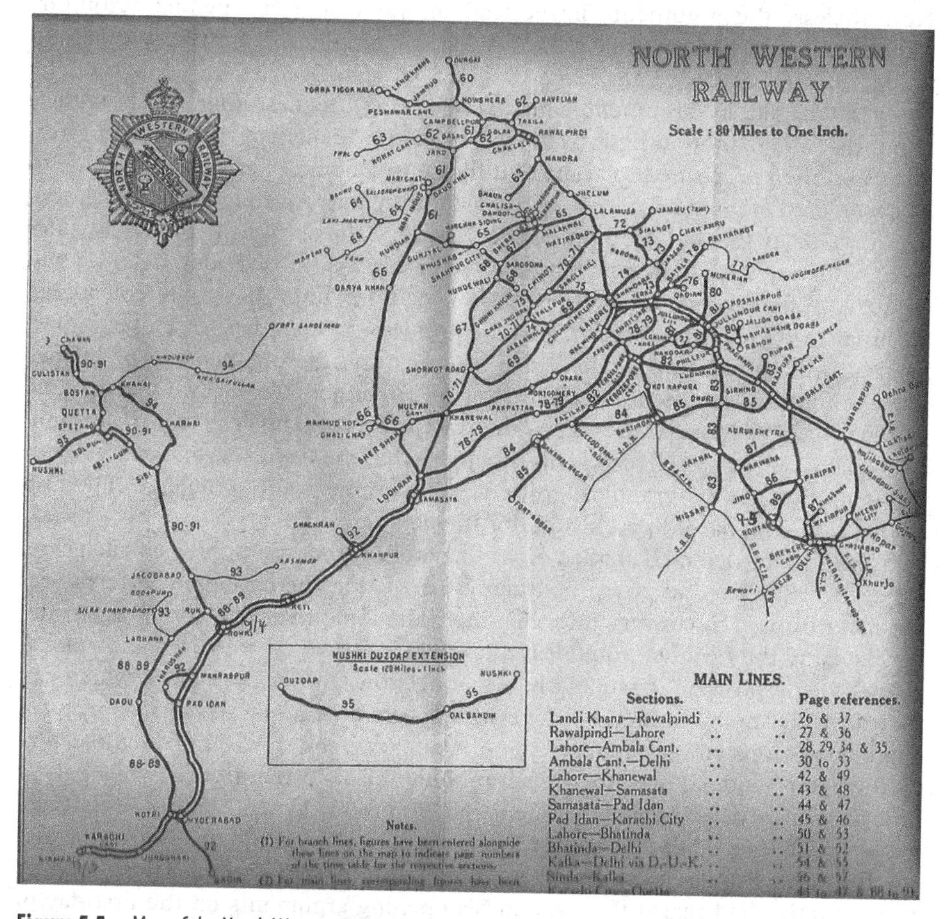

Figure 5.5 Map of the North Western Railway System.
Source: The Papers of Oskar Spate, MS 7886, National Library of Australia.

Spate's Strategic Maps and Issues of Defence

One of Spate's most fascinating contributions to the Punjab Boundary Commission was his attempt to produce what we might call 'strategic maps' – maps that conjured the relative security risks that the proposed Congress-Sikh and Muslim League boundaries might pose in the event of unrest or armed conflict. When the boundary commission hearings began on Monday, 21 July 1947, Spate was in the Lahore courthouse. The proceedings opened with the Congress arguments, and Spate was unimpressed by what he heard. Setalvad, the Parsi lawyer appointed by the Congress to present its case, 'spoke for 3 hours without saying anything which was not either old, obvious, or very dubious', Spate mused in his diary entry from the day (21 July 1947, Spate Papers). Nevertheless, the arguments introduced in the Congress opening testimony inspired a significant amount of the detailed geographical work Spate ended up doing, and which ended up being archived with his papers in Canberra. The Congress began its argument with a (very long) request that the boundary commission give more weight to the 'other factors' clause in the terms of reference, and then proceeded to argue that the census figures were too inaccurate to be considered fit for purpose in a legal hearing over state sovereignty and territorial jurisdiction. At the end of the morning session, Setalvad finished his first statement with an appeal to the boundary commission to consider defence and strategy, an issue which the Muslim League had not, until that point, considered essential to either side's arguments: 'a frontier or a boundary line must, apart from furnishing a natural barrier, be so drawn as, considering the items of defence like communications, is not open to attack from either side easily. Of course, all this is subject to local conditions of the tract which you must take into account' (Sadullah, 1983, p. 34). Despite his insistence that the Congress line of argument rested on dubious geographical grounds, this comment about defence ended up inspiring quite a bit of Spate's work for the next few days.

Spate's maps depicted shaded areas forty miles from the border on each side, for each of the two proposed boundaries, depicting the hypothetical territorial advances from each country into the other. The Congress-Sikh claim, according to Spate and the League, would leave Pakistan in a severely impaired and vulnerable strategic position, particularly in connection with railway lines, which ran perpendicular to the border, for the railways were potential warfronts, a means of deploying troops and accessing remote territory to the west. The League claim, on the other hand, allowed for greater mobility for Pakistani troops, while having little impact on India's ability to defend itself from the opposite side. The strategic maps, which showed the potential movement of the borders in the event of armed conflict in the Punjab, were drawn and presented primarily because the Congress representative had raised the issue in his opening arguments on the first day of the commission. As Spate wrote in his diary, 'At 1215 [Setalvad] finished brief prelim. survey of physical features wch so far as facts in it went was brief; I could have said twice as much on this matter in 15 minutes. However, he raised a strategic aspect which gave me an idea. . .' Spate spent the rest of the afternoon drafting two strategic maps (Figure 5.6). They were probably the maps he was most proud of, and he used them to present an argument which he believed would 'completely demolish that side of the Congress' case' (21 July 1947, Spate Papers). Zafrullah Khan was 'very impressed' by these maps, 'which indeed present an irrefutable argument very graphically', and Spate's argument was incorporated into the Muslim League claim (22 July 1947, Spate Papers).

Zafrullah Khan presented Spate's strategic map to the boundary commission on Monday 29 July. Noting that the Congress representative had introduced the issue of defence during his arguments with a reference to a quote from Thomas Holdich, Zafrullah Khan pointed out the strategic asymmetry between the two future states. As with most of the technical aspects of the Partition, from a strategic perspective, Pakistan was also at a clear disadvantage. Zafrullah Khan noted that the Congress claim would leave Pakistan with very little of the Punjab's developed territory:

> they propose to leave Pakistan a narrow strip beginning with Karachi and ending with Khyber, narrower still in this sense that it is only a very small portion of this strip that can be effectively developed. The rest is desert and hills of no practical purpose with regard to the defence against India. (Sadullah, 1983, p. 314)

Spate believed that the proposed Congress line would leave Pakistan open to invasion by India, and he believed his maps would prove 'devastating' to the Congress claim, at least if strategic considerations were deemed significant. His map showed the two proposed boundary lines, along with the main railway lines, with shaded sections representing the areas within 40 miles of each side of the two borders. As Zafrullah Khan told the boundary commission, 'its [sic] design indicates what would happen if in the case of declaration of hostilities or in case of commencement of hostilities either side advanced forty miles into the territory of the other' (Sadullah, 1983, p. 315).

The most important conclusion that Spate came to regarding defence was the simple fact that the Congress-Sikh line would 'make the defence of Pakistan a hopeless proposition'; this would be so even if the proposed Pakistan and India 'were powers of roughly equal [military] strength', which they were not (Note on strategic aspects of proposed boundaries, Spate Papers, p. 3). This is one of the main considerations that Spate had in mind when he noted (above) that a border had to be 'workable': it had to 'reduce frontier friction', as he put it in his diary. But this was not all that he meant by the expression. Spate wrote in his notes that, with a 'workable boundary' in mind, 'there is precious little to be said in favour of this [the Congress-Sikh] line, even were it a good line on general communal and economic principles, which it most assuredly is not' (Note on strategic aspects of proposed boundaries, Spate Papers, p. 3). The proposed boundary left the area

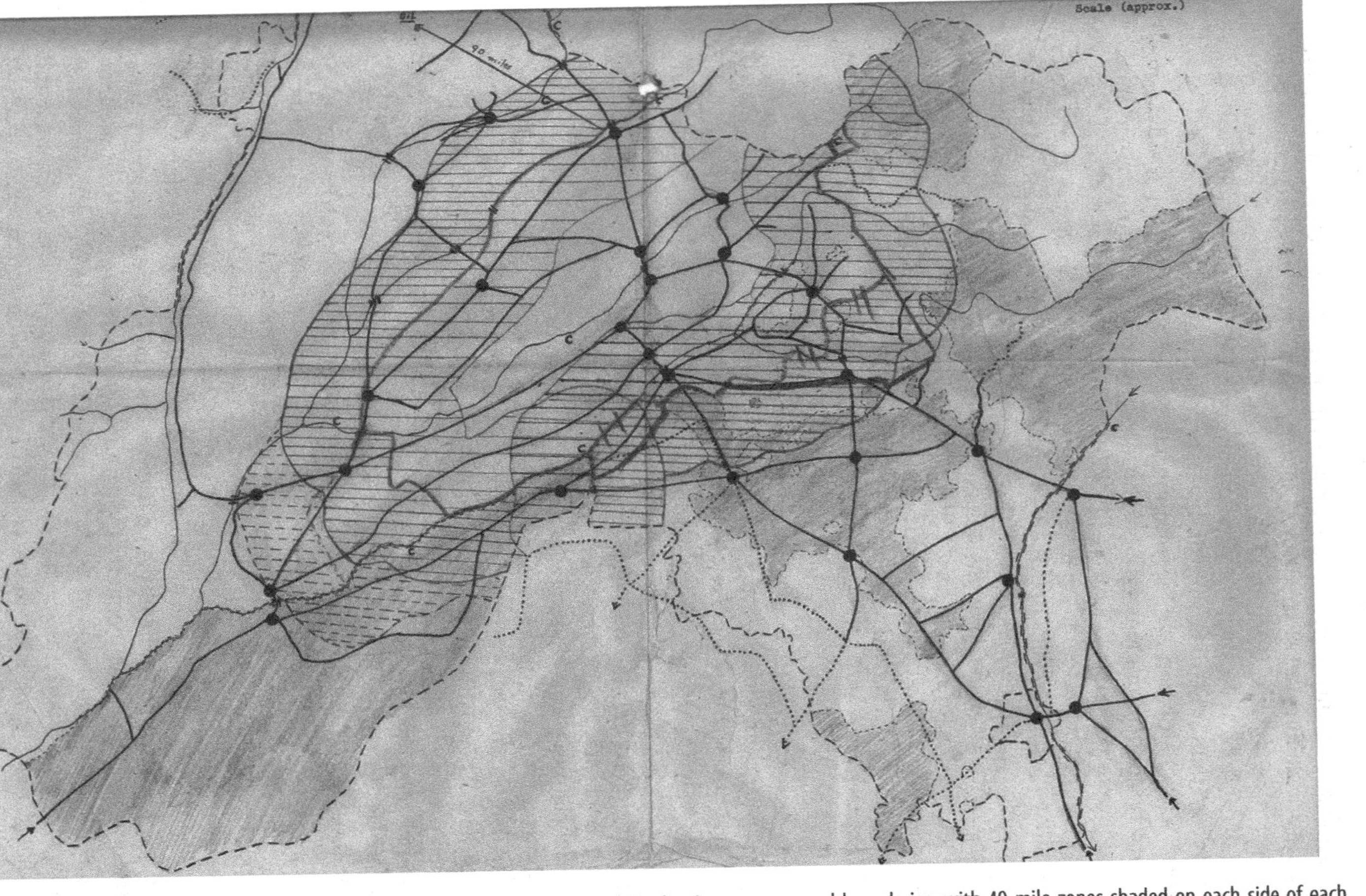

Figure 5.6 Spate's Draft Strategic Map, showing both Congress-Sikh and Muslim League proposed boundaries, with 40-mile zones shaded on each side of each boundary.

Source: The Papers of Oskar Spate, MS 7886, National Library of Australia.

between the border and the mountains with very little in the way of resources, transportation and communication links, and other infrastructure. The area left was simply a 'narrow strip' that was bordered on the 'wrong' side by mountain ranges. Such a division would have left Pakistan without a reliable and well-connected railway system between the potential warfront and Karachi, with the railway line running parallel to the front for a significant distance. Additionally, very few natural resource industries would be left in Pakistan. India, meanwhile, would face very few strategic challenges. The League's proposed boundary, on the other hand, 'would hardly affect the defence of India and might well indeed be a better line for the defence of India than one on the Chenab' (Note on strategic aspects of proposed boundaries, Spate Papers, p. 1). India would still have an overall advantage, but Pakistan would be in a far more manageable position, with more developed territory and greater access to the transportation infrastructure.

With hindsight, it is clear that this style of territorial warfare is not how India and Pakistan have engaged in conflict since independence. Nor is it the way in which other international conflicts have been (and are still) carried out. But at the end of the Second World War, the search for strategic and defensive borders was a vital part of bordering processes, of how politicians involved in these processes thought and acted, and how geographers became implicated in them. Zafrullah Khan presented these arguments about strategy and defence along with Spate's maps to the Punjab Boundary Commission, a clear example of Spate's geographical and technical knowledge being inserted into the political and judicial proceedings.

Visible Boundaries

Spate's geographical analyses also helped the Muslim League delegation devise their rebuttals and counter-claims during the Punjab Boundary Commission hearings. The Congress-Sikh claim argued that central Punjab was a single economic and physical unit, incapable of being partitioned effectively; they used this statement to claim the entirety of the central belt. Spate argued that 'the Punjab *proper* – i.e. the land between the five rivers – is such an integer that any division of it cannot fail to inflict serious economic damage on both parties'. He continued, 'in fact, the Punjab can be considered as a central block – the triangle between Jhelum and Sutlej – with two wings: Sind Sagar Doab and Trans-Indus to the West, and Delhi Doab (Sutlej-Jumna) to East'. The unity of the central block, as Spate saw it, had a great deal to do with the canal system. The Congress-Sikh claim, he said, actually cut 'right through the integrated canal area and split it hopelessly'. The Muslim League claim, on the other hand, 'most nearly preserved the division which an overall geographical view would suggest'. He concluded, 'A broad geographical survey therefore favours the Muslim case more than the non-Muslim' (Note on some geographical arguments adduced by Congress, Spate Papers).

Spate also critiqued the Congress delegation's use of colonial boundary-making expertise. In his oral arguments, Setalvad cited Thomas Holdich as an expert in boundary-making in order to argue that natural geographical features made ideal boundaries, and that borders were meant to be strategic 'barriers' between nation-states. Holdich had argued that one of the primary purposes of a border was to deter and limit aggression, and that 'natural' boundaries (rivers, mountains and buffer states, an anachronistic and irrelevant option by 1947) were the most effective way to secure a border. Setalvad had drawn on Holdich to argue that the boundary should, when possible, be based on physical geographical features and should discourage aggression on both sides. However, Spate argued that Setalvad's application of Holdich was relatively superficial suggesting that he was more interested in mobilising the intellectual authority of one of the British Empire's most celebrated geographers in support of the Congress delegation's political argument, rather than presenting a more geographically focused (and, for Spate, rigorous) argument. Spate noted that Holdich 'had undoubtedly much experience of delimitation and demarcation but his general views on frontiers are not now widely accepted'. However, Holdich had noted that physical boundaries were not necessarily the most effective, and that borders are made effective by a combination of other factors. Spate's definition of a boundary was more nuanced and complex than Holdich's, of course; Holdich dealt with territorial frontiers and the limits of colonial control in an age of empire and imperial security. By 1947, borders between European nations, both in Europe and in the colonies, had changed dramatically, and the work of boundary-making, according to Spate, required more knowledge of the populations which would be controlled by new boundaries. A 'good boundary', Spate opined, '1/ should be easily identifiable on the ground by anybody'. Physical boundary markers and natural boundaries could often serve this purpose, and rivers were particularly suitable due to their visibility, but, he continued, 'it should also be a boundary administratively workable with the least friction'. And, finally, a boundary 'should so far as possible avoid economic dislocation' (Note on some geographical arguments adduced by Congress, Spate Papers).

The Congress-Sikh lines did not fulfil either of these criteria. Spate said it was 'a highly incompetent piece of drafting, a lazy line making use of existing local boundaries without any reference whatever as to whether they are suitable as inter-state boundaries'. Setting aside the gerrymandered and geographically inaccurate Congress red map, and the debatable claims it made to contiguous areas, the Congress-Sikh line itself was, in Spate's opinion, 'very fallacious' and applied 'no knowledge of the immense body of fact and argument which modern geography has rendered available and relevant to problems of this nature'. Indeed, the Congress line did not even fulfil those requirements that Holdich had set out. This is not terribly surprising, given that the partitioning process was less interested in creating the boundary itself and more interested in the political negotiations over which nations could claim sovereignty and sections of territory. Spate wrote, 'the non-Muslim line violates their own principles and those of "natural boundaries"' (Note on some geographical arguments adduced by Congress, Spate Papers).

Spate also saw the issue of rivers as important and multi-faceted, and requiring a knowledge of physical geography. Indian rivers were well-known for their fluctuating and inconsistent fluvial dynamics. Prone to dramatic shifts in course, level and flow, fluctuations in current and discharge, and spatial and temporal variation in erosion rates, rivers in India were not particularly suitable as physical boundaries. In his 'Note on Some Geographical Arguments Adduced by Congress', which he prepared for Zafrullah Khan and his team, Spate gave a geographical explanation for why Indian rivers would not make a good boundary in this case: 'they are constantly shifting their courses, eroding and alleviating, so that a boundary fixed in a river would either have to be changed with the river or form a series of loops on each side of it – as can be seen on any of the District boundaries where formed by rivers'. Additionally, the Punjab rivers were 'used for irrigation and timber-rafting and may be used for hydro-electric power, thus giving probable sources of friction over abstraction of water, access to banks and islands, etc.'. For Spate, 'the only sort of river which makes a really good boundary is one flowing in a deep rocky canyon, crossed at only a few points, and never shifting their courses, and useless either for navigation or rafting, irrigation or hydro-electric power. Few rivers have these distinctly negative virtues, and the Punjab streams are certainly not of them'. However, the rivers in the Punjab were not without their value as potential boundaries. Spate returned to the importance of the visibility of the boundary regularly in his notes. Rivers were a particularly good visible boundary, 'easy for the most illiterate to know and to see'. Canals, roads, and railways all shared this same feature, and had the additional feature of remaining fixed. 'The disadvantage [was] that both sides have normally a common interest in the use of them'. Of these, Spate argued that railways were the best option if a pre-existing 'man-made line on the land' were to be taken as part of a boundary. He said, 'Railways. . . are under administrative surveillance and traffic can only join at fixed points'. The only disadvantage was 'that it may thereby become strategically useless' (Note on some geographical arguments adduced by Congress, Spate Papers).

The debate over the effectiveness of 'natural' versus 'man-made' boundaries speaks to wider issues in boundary-making regarding the ways in which nature and space are produced and ordered for the purpose of creating 'working' borders. Spate's assertion that the best river boundaries are those that run through canyons, have very few crossings and are nearly 'useless', indicates that 'natural' boundaries are only effective if they are very difficult to overcome, either by crossing them or manipulating them, and are 'untouched' by human activity. In this 'nature-as-barrier' perspective, 'nature' is deemed strong enough to enact and enforce the territorial will of the nation-states which it demarcates only if it has proven 'strong' enough to withstand man's attempts to 'tame' and 'control' it. The visibility of the boundary, therefore, is tied up with its accessibility and its permeability. It is not enough, for Spate, that Punjabis could see and recognise the River Sutlej, or the River Chenab, as a boundary; these rivers had been 'harnessed', so to speak, and put to work producing energy and irrigation channels. They were not barriers, they were infrastructure, mobilised by the colonial government to unite, rather than divide, the Punjab. Importantly, this same issue characterised the possible use of roads and railways, that 'both sides have normally a common interest in the use of them'. Yet man-made boundaries were fixed in space, and visible and knowable to the population. In this sense, boundaries needed to be visible in order to make the territorial limits of the new states known and enforced. The visibility of the new borders was especially important in the Punjab and Bengal because they dictated a new spatial order: two states to be carved from one.

Deconstructing the Red Line

Spate picked apart the Congress delegation's method for producing the red map and the disputed Lahore city population figures. When it came to the population data, as he put it, 'the Congress and the Sikh memoranda make a great deal of play with the unreliability of the 1941 Census in Lahore City' (Note on the Population of Lahore City, Spate Papers). This was the case. The Congress and Sikh claims relied heavily on the notion that the 1941 census data were inaccurate, primarily due to the politicisation of the census by communal politicians who sought to use the census to gain power and influence. As a result, the 1931 census data was also used by the parties in determining majority populations. As I noted above, such political manipulation surrounding the census had occurred for decades, and particularly since the implementation of separate electorates for Hindus and Muslims in 1909 as part of the Morley-Minto reforms. Spate believed that the census data was certainly not accurate to any ideal degree, but he suspected that the data was likely skewed in favour of the non-Muslim side if the data was as inaccurate as the Congress memorandum claimed, contrary to the Congress arguments. Indeed, he wrote in his notes, 'this unreliability may be admitted without the conclusions they wish to draw following' (Note on the Population of Lahore City, Spate Papers). The Congress claim stressed the importance of Lahore for India, and the oral and written arguments showing the Lahore city census figures were part of the Congress bid for the city. Radcliffe was unlikely to award both Lahore and Amritsar to India, but that did not stop the Congress delegation from arguing for both cities; in fact, according to some historians, Nehru and the Congress leadership were surprised when Lahore went to Pakistan.

The Congress red map was the most pronounced example of gerrymandering presented to the Punjab Boundary Commission. The red map attempted to deploy the authority of the cartographic gaze by abstracting and simplifying the geographic and statistical data of the census records and survey maps, and by using the seemingly scientific authority of a map to obfuscate that data. In doing so, the red map challenged the geographical accuracy of the Muslim League maps, despite the fact that, as Spate observed, 'a ve[r]y [sic] cursory inspection of the new Congress map shows that it cannot be accurate' (Note on the Congress Map, Spate Papers). Spate placed the 'accuracy' of his geographical thinking and maps at the political disposal of the Muslim League, and in his mind for rational rather than simply partisan ends. The map claimed for India large areas of what were clearly Muslim-majority *tehsils*. In order for these claims to be accurate,

> the Muslims must be (a) almost equal to non-Muslims in the non-Muslim areas, (b) in overwhelming majority in the Muslim areas, which areas (c) must be more densely populated than the rest. A distribution like (a) and (b) is inherently extremely unlikely, one might almost say impossible, and a glance at a topographical map shows (c) to be absurd. (Note on the Congress Map, Spate Papers)

Spate wrote in his notes, 'On general grounds then anyone with any experience of dealing with population maps would be justified in suspecting very strongly the authenticity of the map'. Spate drew a population map (Figure 5.7) of Sheikhupura *tehsil*, chosen at random and located in the middle of the disputed area of central Punjab, between the Ravi and Chenab rivers, to illustrate his point.

On the Congress map, much of Sheikhupura *tehsil* had been shaded red and claimed for India. Spate found the claim fallacious:

> A rough population map was constructed from which it seemed fairly clear that the method used in constructing the Congress map was (i) to subtract obvious Muslim pockets (ii) to regard these as being practically solid Muslim (iii) deduct some sort of figure for these Muslims large enough to make the new total non-Muslim appear in majority over the rest of the area. (Note on the Congress Map, Spate Papers)

In other words, the Congress map assumed that the Muslim populations of each district and *tehsil* were heavily concentrated in a few small areas, thereby supporting a claim of contiguity in terms of land area. According to Spate, 'the density of population of the Muslim areas would have to be at least 25% higher than that of the area as a whole for this method to work; such an assumption is of course glaringly false' (Note on the Congress Map, Spate Papers). In fact, Spate's Sheikhupura *tehsil* map (Figure 5.7) 'shows that the population is in fact spread out pretty evenly, as might be expected'. In other words, the map misrepresented the spatial distribution of the populations, moving Muslim majorities into smaller spatial units in order to make a false claim to contiguity.

On the last day of the proceedings in Lahore, Setalvad confirmed Spate's calculations: 'As I had deduced on general grounds', Spate wrote in his diary, 'the Muslim "pockets" have whacking great majorities and the big red areas very thin ones' (31 July 1947, Spate Papers). What Spate does not say, of course, is that the

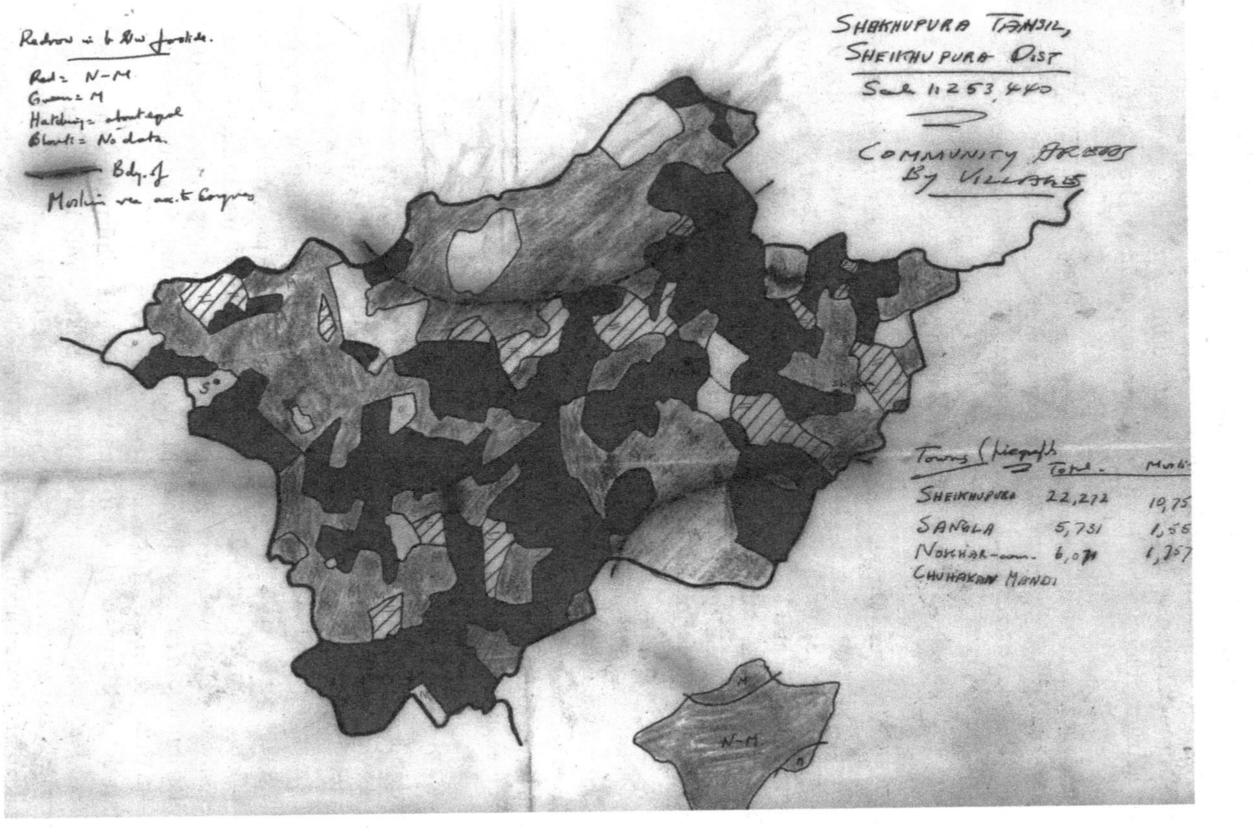

Figure 5.7 Draft Map by Spate of Sheikhupura *Tehsil*. Showing Muslim and non-Muslim majority villages, with Congress-Sikh claim from red map shown for reference below. The map indicates how the Congress red map tried to claim contiguity by arguing that Muslim majorities were contained in small 'pockets', and therefore able to be included in East Punjab as part of larger, contiguous non-Muslim areas. The map above, drawn by Spate, shows that, in fact, Muslim-majority and non-Muslim-majority villages were evenly dispersed, and represented inaccurately on the red map.

official League map might have captured a more accurate picture of population distribution, but his draft map of Sheikhupura *tehsil* is a distilled picture of the fundamental issue of scale and the Punjab population. A boundary could not reasonably be determined using such a small scale because the populations were not contiguous at these more local levels. Indeed, the very idea of contiguity breaks down at this level because local life involves everyday mobility and circulation. Ultimately, while Spate's draft map succeeds in providing a geographically sound rebuttal to the geographically dubious Congress territorial claim, it also depicts the limits (which he himself knew) of population mapping as a scientific or rational mechanism for determining a boundary based on majority.

Assessing the Sikh Claims

The Sikh claim presented an intractable problem for the Punjab Boundary Commission, as well as for the Muslim League's own claims and rebuttals. While this study focuses on the Muslim League's geographical imaginings and therefore does not examine in detail the role of the Sikhs specifically with regard to the politics of Partition, the Sikh claim to the Punjab central belt became central to the Congress strategy because of the location of the majority of their community. Spate's archive therefore engages with the Sikh claims, and illustrates how geography not only underpinned the initial claims but provided a set of tools for refuting or challenging the claims of the other parties.

Spate found the Sikh demands frustrating and unrealistic, arguing the Sikhs were played as 'catspaw' by the Congress for political expediency in countering the League (Spate, personal diaries, 19 July 1947). In the early 1940s, a separatist Sikh independence movement began to gain support, calling for a state called Khalistan, located in the centre of the Punjab, with its capital at Amritsar. The total Sikh population in India was, at the all-India scale, minimal, especially when compared with the Hindu and Muslim populations; the Sikhs were unique because the Punjab was, and is, the heart of their religious, cultural and linguistic community. The Golden Temple in Amritsar, which would later become the site of a battle between militant Sikhs and the Indian state in 1984, remains the centre of Sikh religious life.

The Sikh claim was based primarily on historically constructed affective ties to the Punjab – pointing once more to those expansive and fluid conceptions of territoriality, in this instance as homeland and sacralised space, that Spate pointed to in his RGS lecture. The Sikhs had a majority only in Amritsar and Ferozepur districts, and so their claim necessarily relied purely on the 'other factors' clause in the boundary commission terms of reference. They presented a list of (over 700) sacred sites located in the Punjab which they insisted go to India. The Sikh representative also built a set of oral arguments by appealing to a historical-communal tension between Indian Muslims and Sikhs, focused primarily on atrocities committed by Muslims upon Sikhs, which precluded those groups from living together.

The narrative drew primarily on instances of Muslim-Sikh conflict in the Punjab during the Mughal and Sikh dynasties, before the British had annexed the Punjab in 1849. Unsurprisingly, Zafrullah Khan, supporting the Muslim League, challenged this historical account by providing an alternative one, which downplayed the role of community in historical conflicts between Muslims and Sikhs, casting historical violent incidents as being between individuals. Additionally, the Sikh claim stressed, at great length, the unique relationship that Sikhs had to the land of the Punjab, arguing regularly that the Sikhs were 'rooted in the soil', a phrase that Spate took issue with: 'A fantastic argument; when are you "rooted in the soil"?' (Comments on the Sikh Case, Spate Papers). Spate found this claim false primarily because of its subjective, malleable and ultimately affective meanings, even while elsewhere alluding to the importance of 'other factors', some of which were 'subjective' factors. The Sikh representatives' claim to being 'rooted in the soil' stressed the importance of agricultural livelihoods among Sikhs, calling all others (primarily Muslims) in the Punjab 'floating' populations, able to migrate easily and with a less valid economic, historical or cultural claim to territory, a claim which Spate called 'ridiculous' (Comments on the Sikh Case, Spate Papers). Spate believed that any of the communities in the Punjab could mobilise an argument using historical and economic ties to the productivity and cultural significance of the territory. This argument was, presumably, too nebulous and difficult to defend with geographical evidence – a different form of vagueness.

This preference for seemingly rational and reasoned evidence is filtered through much of Spate's commentary and observations, and all sides (including the official British side) were subjected to a similar critique. In his notes on the Sikh case, which he addresses point by point, he calls attention to arguments that are unconvincing in geographical or statistical terms, using phrases like, 'mainly irrelevant', 'grossly irrelevant', 'all of this is half-truth', 'these observations by themselves prove nothing to the point', 'there is nothing in this argument', 'this is a mere fiat', 'this is really rich' (Comments on the Sikh Case, Spate Papers). Unlike the Sikh sentiment of being 'rooted in the soil', population statistics were, according to Spate (and others, including Stephen Jones, discussed in Chapter 2) a more useful tool for the purposes of a territorial partition. Unfortunately, the overall numbers were too small for the Sikh delegation to stake any claim based on population; the community was located directly in the middle of one of the two regions to be partitioned, and these affective arguments in favour of cultural, historical and economic ties to the land were the only hope the Sikhs had at gaining any of their wishes.

From Lahore to Simla

Spate used the Sikh claim, based primarily on 'other factors', to illustrate just what he believed 'other factors' meant: 'I would suggest that the natural meaning, in accordance with precedent usage, is that it is analogous to the powers rightly given to boundary demarcators in the field to make minor deviations from the originally defined line on paper to meet obvious geographical anomalies, being guided by common sense. . .This is such obvious common sense that I hardly see how any other view could occur to rational men' (Comments on the Sikh Case, Spate Papers). Spate exhibited the same bias towards European rationality and reason described above, and at length in the previous chapters. As postcolonial observers writing nearly 70 years after Partition, we are often hesitant to make such bold statements in support of 'fact' and 'common sense' with regard to the process, and rightly so in light of the violence and trauma that attended the weeks and months following Partition. Spate, however, was 'in the thick of it', as he said, and worked as an ambivalent yet complicit participant in the drama of a crumbling empire (1947, p. 202). He was genuinely concerned with the real, life-and-death implications of Radcliffe's pending decision, and genuinely believed that reason and the search for geographical order were vital to the attempt to make sense of the complicated and often confusing political demands and arguments which characterised the process for transferring power in India.

Indeed, many scholars and activists have appealed to a similar rationality as a means of understanding the process, and of combating the kinds of violence and religious fundamentalism which have developed in parts of the subcontinent since 1947. And yet the Congress and Sikh claims, which were based on very different, political and affective arguments and assumptions about the 'other factors' clause, were still filtered through this rational discourse which was rooted in a philosophy of colonial scientific reason. The ambiguity of the geographical terminology applied to so much of the process allowed the terms of reference to be appropriated in political terms while retaining a veneer of scientific rationality. This allowed for the presentation of two vastly different arguments, both of which were given the same political weight. The Congress-Sikh claims were certainly not accurate in geographical terms but they were still valid political claims, in the sense that they represented, in legal discourse, a proposed boundary which the Congress delegation could use as a bargaining tool.

During his last few days in India, Spate went to Simla where the boundary commission was deliberating under Radcliffe's leadership ('the Ahmadis thought I should be where the action was', he noted in his diary) (Spate, 1991, p. 58). Simla, a hill station that became the preferred summer retreat of the British in India, was transformed by the imminent transfer of power. Simla, and other hill stations like it, had been built 'by the British, for the British', in the words of Kathy Baker (2009, p. 3). The hill stations were retreats, designed to allow the British in the colony to escape from the 'real' India, from its weather, its dirt and its people. Located at the base of the Himalayas, Simla was sparsely populated when the British first arrived there in the early 19th century, and, as the climate was generally preferable to British colonists and there was little in the way of infrastructure, the hill stations were developed in ways which would make them familiar and comfortable to the British settlers who visited and lived there. This included the planting of gardens which evoked the gardens of England, building houses and cottages in the Victorian styles reminiscent of 'home'. By the turn of the century, Simla was the summer office of the Viceroy of India and became one of the most important sites of British power in India. The importance of Simla as the site from which Radcliffe published his final award does much to discredit Mountbatten's (and the colonial regime's) insistence on Partition as an Indian, rather than colonial, process. It is not surprising that Radcliffe and the boundary commissions retreated to Simla for their final deliberations, away from the political and cultural centres of Calcutta and Lahore where the Indian and Pakistani nationalists had had their say, and into the space of British governance. In the final hours, the moments that counted the most, the decision was made at one of the most important sites, both symbolically and materially, of the Raj.

Conclusion

By 10 August, Spate had returned to Karachi for the Pakistan independence celebrations, which 'were on a modest scale, almost half-hearted', he thought. The new leaders 'were too busy desperately improvising a government from next to nothing to have much time for celebrations'. Spate remained in Karachi, and on Independence Day, which was celebrated on 14 August in Pakistan, 'spirits were high' and 'there were crowds and cheers as Jinnah and Mountbatten drove by in state' (Spate, 1991, p. 59). After Independence, Spate returned to London, where he was invited to address the Royal Geographical Society on the boundary dispute. On 8 December 1947, he presented a paper titled 'The Partition of the Punjab and Bengal' (Spate, 1947). Decades later, he drew on the contents of this paper, along with his journals, to write his memoir, *On the Margins of History: From the Punjab to Fiji*, published in 1991. Much of the content of these academic pieces mirrors the observations and arguments he recorded during his time in the Punjab, but one can see a subtle change in Spate's geographical discourse as he moves from the political space of the Lahore courtroom to the academic and professional space of the RGS.

There is a wider paradox inherent in the roles and productive outputs of geographers (a paradox which is similarly apparent in other social science disciplines like anthropology and psychology). Academic and scientific ideals of rationality, objectivity and neutrality are very often at odds with the kinds of highly charged, high-stake contexts in which geographers often find themselves. Boundary-making negotiations are a prime example of just such a politically charged situation. Spate's role in the boundary-making process in the Punjab exemplified this paradox and was in fact made all the more problematic given the context: the transition to postcolonial nation-states. In the courtroom, Spate was required to make technical assessments and recommendations which would have a direct influence on a legal process and set of political judgements, while in the lecture theatre, he was required to maintain a detached and objective perspective for the purposes of academic interpretation and instruction. Politics became a tricky subject in this context: central to the boundary-making process but viewed with suspicion, and translated unevenly for the detached and scientific setting of the RGS.

Spate's geographical expertise did, however, carry a certain amount of intellectual weight, both at the time and after Independence. Some historical geographers have noted the ways in which Indigenous and imperial discourses regarding territory, property, and sovereignty came into contact in legal and academic spaces. Orlove (1991), Sparke (2005), Burnett (2000), Wainwright and Robertson (2003) and Belmessous (2012) have written about the tendency of colonial officials and colonial courts to ignore, trivialise or distort Indigenous forms of spatial and political knowledge in order to claim territory, resources and sovereignty. In India, however, the nationalist parties and the British government shared a complex politico-territorial discourse through which the official boundary dispute was conducted. Spate was able to evaluate the various party claims presented to the Punjab Boundary Commission primarily because the main parties constructed their claims using a shared language of territorial sovereignty. Spate could argue, for example, that Sikh claims of having been 'rooted in the soil' were not legitimate according to geographical standards because that kind of historical territorial relationship was a recognised geographical claim in other contexts. Similarly, the Congress red map could be denounced as a gerrymandered land grab because the map was created within the shared discourse of statistical mapping. Unlike other colonial contexts, where European discourses of territoriality and geographical knowledge rendered Indigenous knowledge invisible or illegitimate, the discourse of the boundary commissions sat firmly within the 'quasi-judicial', as Spate called it, and territorio-cartographic discourses of state and treaty-making recognised by European governments and courts of law (1947, p. 205). The legacies of this discourse are, of course, ongoing and contested, and highlight the ways in which postcolonial nation-states in the subcontinent have inherited and appropriated colonial institutions and modes of knowledge in their forms of governance and control.

Spate began his presentation at the RGS by addressing this conundrum:

> The term 'political geography', which we all allow, must include the influence of politics on geography as well as the converse; and in this case, indeed, political considerations largely conditioned new geographical arrangements. What is a mere exercise in applied geography to us sitting in this hall was not only a matter of political life and death to the parties involved but literally of life and death to thousands of men, women, and even children. However dreadful in its working, the political factor is fundamental. (Spate, 1947, p. 201)

Spate, even as an ambivalent participant, was able to recall the violent and menacing absurdity of empire, calling attention to the tangled complicities of the

British expert in the context of the breakup of empire. Spate's response was often, particularly in his personal diaries and memoirs, to adopt a tone of irony in order to criticise and even to make sense of the material and epistemic violence of empire. Spate also noticed and anticipated the ways in which his involvement in the boundary commission hearings might be considered to have been 'political' by the RGS establishment, arguing (somewhat contradictorily) that he was less politically motivated than the circumstances might appear, but that he was, indeed, drawn into and absorbed by a politically charged atmosphere of 'life and death'.

As 'an expert witness' employed by the Muslim League, Spate 'was bound to support the interpretation more favourable' to the League. As I have noted previously, however, he 'did not need to exercise any licence' because 'the Muslim case seemed [to him] entirely legitimate' (Spate, 1947, p. 201). Even so, he recognised the contradiction for the detached expert applying geography during political processes, saying

> while I have been at pains to be impartial in the sense of resolute adherence to fact, and have striven to avoid bias in interpretation so far as I can, I do not pretend to a completely academic detachment. I was not above the battle but in the thick of it, and being a political animal I thoroughly enjoyed it. (Spate, 1947, p. 202)

Nevertheless, he generally approached both his job with the Muslim League and his academic output after Partition with a discourse of objectivity and distance.

References

Ahmad, N. (1968). *An Economic Geography of East Pakistan*, 2nd ed. London: Oxford University Press.

Ahmed, A.A. (2010). The Paradoxes of Ahmadiyya Identity: Legal Appropriation of Muslim-ness and the Construction of Ahmadiyya Difference. In *Beyond Crisis: Re-evaluating Pakistan*, edited by N. Khan, 273–314. New Delhi: Routledge India.

Baker, K. (2009). The Changing Tourist Gaze in India's Hill Stations: From the Early Nineteenth Century to the Present (Working Paper No. 25). *Environment, Politics and Development Working Paper Series*. London: Department of Geography, Kings College London.

Bayly, S. (2009). Conceptualizing Resistance and Revolution in Vietnam: Paul Mus' Understanding of Colonialism in Crisis. *Journal of Vietnamese Studies* 4, 192–205. https://doi.org/10.1525/vs.2009.4.1.192.

Belmessous, S. (2012). *Native Claims: Indigenous Law Against Empire, 1500–1920*. Oxford: Oxford University Press.

Brotton, J. (2012). *A History of the World in Twelve Maps*. London: Penguin UK.

Burnett, D.G. (2000). *Masters of All They Surveyed: Exploration, Geography, and a British El Dorado*. Chicago: University of Chicago Press.

Chapman, G.P. (2012). *The Geopolitics of South Asia: From Early Empires to the Nuclear Age*. Surrey, UK: Ashgate Publishing, Ltd.

Chatterji, J. (2007). *The Spoils of Partition: Bengal and India, 1947–1967*. Cambridge: Cambridge University Press.

Cheema, P.I. (2000). *The Politics of the Punjab Boundary Award*. Heidelberg: Heidelberg Papers in South Asian and Comparative Politics.

Chester, L. (2009). *Borders and Conflict in South Asia: the Radcliffe Boundary Commission and the Partition of Punjab*. Manchester: Manchester University Press.

Copland, I. (1991). The Princely States, the Muslim League, and the Partition of India in 1947. *The International History Review* 13, 38–69. https://doi.org/10.1080/07075332.1991.9640572.

Devji, F. (2013). *Muslim Zion: Pakistan as a Political Idea*. Cambridge, Massachusetts: Harvard University Press.

East, W.G., Spate, O.H.K., and Fisher, C.A. (1971). *The Changing Map of Asia: A Political Geography*. London: Methuen.

Elden, S. (2013). *The Birth of Territory*. Chicago: University of Chicago Press.

Hamza, El. (1942). *Pakistan, a Nation*. Lahore: Sh. Muhammad Ashraf.

Jalal, A. (1994). *The Sole Spokesman: Jinnah, the Muslim League and the Demand for Pakistan*. Cambridge: Cambridge University Press.

Jones, K.W. (1989). *Arya Dharm: Hindu Consciousness in 19th-Century Punjab*. Delhi: Manohar.

Kamran, T. (2016). Mapping the Politics of Ahmadi Exclusion and the Rise of Khatam-i-Nubuwwat in Pakistan. In *Democratic Transition and Security in Pakistan*, edited by S. Gregory, 21. Abingdon: Routledge.

McDermott, R.F., Gordon, L.A., Embree, A.T., et al. eds. (2014). *Sources of Indian Tradition: Modern India, Pakistan, and Bangladesh*. New York: Columbia University Press.

Orlove, B.S. (1991). Mapping Reeds and Reading Maps: The Politics of Representation in Lake Titicaca. *American Ethnologist* 18, 3–38.

Sadullah, M.M. (1983). *The Partition of the Punjab, 1947: A Compilation of Official Documents*. Lahore: National Documentation Centre.

Saeed, S. (2018). Secularization of Politics: Muslim Nationalism and Sectarian Conflict in South Asia. In *Tolerance, Secularization and Democratic Politics in South Asia*, edited by H. Iqtidar and T. Sarkar, 50–74. Cambridge: Cambridge University Press.

van Schendel, W. (2005). *The Bengal Borderland: Beyond State and Nation in South Asia*. London: Anthem Press.

Sparke, M. (2005). *In the Space of Theory: Postfoundational Geographies of the Nation-state*. Minneapolis: University of Minnesota Press.

Spate, O.H.K. (1943). Geographical Aspects of the Pakistan Scheme. *The Geographical Journal* 102, 125–136. https://doi.org/10.2307/1788890.

Spate, O.H.K. (1947). The Partition of the Punjab and of Bengal. *The Geographical Journal* 110, 201–218. https://doi.org/10.2307/1789950.

Spate, O.H.K. (1948). The Partition of India and the Prospects of Pakistan. *Geographical Review* 38, 5–29. https://doi.org/10.2307/210736.

Spate, O.H.K. (1991). *On the Margins of History: From the Punjab to Fiji*. Canberra: National Centre for Development Studies, Research School of Pacific Studies, Australian National University.

Spate, O.H.K., Learmonth, A.T.A. and Learmonth, A.M. (1972). *India and Pakistan: Land, People and Economy*. London: Methuen.

Tan, T.Y. and Kudaisya, G. (2002). *The Aftermath of Partition in South Asia*. London: Routledge.

Tayyeb, A. (1966). *Pakistan: A Political Geography*. Oxford: Oxford University Press.

Wainwright, J. and Robertson, M. (2003). Territorialization, Science and the Colonial State: the Case of Highway 55 in Minnesota. *Cultural Geographies* 10, 196–217. https://doi.org/10.1191/1474474003eu269oa.

Chapter Six
Partition to Partitions: New Avenues for Historical Geography

Introduction

Throughout the book, I have presented, through the layering of discrete temporal frames and archival sources, a historical geography of Partition. I have illustrated how Partition can and should be thought about as concerning both geography and history, and space and time; as an event and process; as a line on a map and a division in people's lives; as a new dawn for some and the end of the line for others; and as subject to multiple, competing and shifting interpretations originating from different places, perspectives and positions in social and political hierarchies.

In some ways, the book opens up more questions than it answers. Questions about the simultaneous conflicting pressures of (a perceived) ethical imperative to 'act' and practical and discursive limits placed upon geographers in the context of applied geographical practice; about the productive work of territorial partitions in the context of bordering practices in the postcolonial world; and about the relationship between the historical development of bordering practices and the maintenance of certain global, regional and local distinctions and modes of categorising people and space. 'Partition', in its many theoretical and conceptual guises, continues to animate lively scholarly debate, political and social activism and revisionist projects, working as both an arbiter and object of South Asian modes of remembering, of forgetting, of making and re-making identities, histories and geographies in and beyond the Indian subcontinent.

Mapping Partition: Politics, Territory and the End of Empire in India and Pakistan, First Edition. Hannah Fitzpatrick.

I have sought to make explicit some of the discursive and material connections and entanglements between the geographical knowledge produced and normalised in the late 19th century and the geographical logic which underpinned the eventual Partition of the subcontinent in 1947. I have done so by creatively engaging and mobilising a variety of eclectic sources, produced by both British colonial writers and Indian writers. I have focused on the ways in which geographical knowledge production and applied geographical practice in the context of the later period of the empire in India produced the conditions and limits of spatial and geographical thinking at the heart of the partitioning process in India. I have demonstrated how spatial boundaries and divisions in the Punjab, and in India more broadly, were produced – rather than given – and normalised through the publication and dissemination of geographical materials such as atlases, maps, diagrams, gazetteers, census reports and so forth.

Chapter 2 excavated a selection of texts from this colonial geographical archive, notably William Wilson Hunter's *Imperial Gazetteer of India* and Thomas Holdich's work on boundary-making. Both texts made their way into the Punjab Boundary Commission, at times with greater accuracy than others, and we saw in Chapter 4 how the colonial ideal of cartography and logics of boundary-making were put to work dismantling the very empire they had earlier been used to make. Hunter's gazetteering project imagined itself as a complete geographical account of British India, useful to administrators and educational for the British public. Its later iterations were increasingly polished and the reach of its dissemination grew, but its increasing ubiquity and glossy veneer only hid the cracks that had been apparent from the outset. The gazetteers therefore depict a *colonial geography* of India, including its assumptions, its obfuscations and its aspirations.

Chapter 3 introduced a select set of Indian Muslim intellectuals and politicians, reconsidering some of their most well-known articulations of the two-nation theory and Pakistan in response to and in conversation with the colonial geographies explored in Chapter 2. In particular, this chapter explored what David Scott has called 'the nature of the terrain available for the colonized to produce their responses' (Scott, 1995, p. 197). Historians, particularly those who work on India's intellectual history, have examined many of these ideas in detail, and this chapter engaged recent work on the concept of minority as both a statistical and political category. A geographical analysis of this work shows that minority is also a *spatial category*, one which Indian Muslim thinkers worked to mould in creative ways by using Islam's transnational and mobile geographies to overcome British colonial geography's attempts to fix Indian religious communities onto maps of British India. In this context, Indian Muslim politicians attempted to create a geography of post-independent India that could work in multiple and, to some extent, contradictory ways. The territorially bounded nation-state, with its ideal of a homogeneous population, emerged as the most logical option according to the dominant geographical logic of the state, even though it sat uncomfortably with the imaginative geographies of Indian Muslim thinkers before Partition. Taken together, these two chapters show how geographical knowledge was at work before 1947, producing the cartographic conditions for a territorial partition to be imagined or thought.

In 1947, the techniques and procedures perfected for the consolidation and maintenance of empire explored in Chapter 2 were deployed for the purposes of dismantling that same empire, and for the re-fashioning of a new system of independent nation-states. Chapter 4 therefore moved into the space of the Lahore High Court, where the Punjab Boundary Commission met for 10 days in July and August of 1947, to hear the claims of the main political parties and communities who would be affected by Partition. In this legal venue, the geographical knowledge produced in and of colonial India was politicised in particular ways, rendered as evidence for the purpose of convincing judges of their validity. Claims based on colonial geography sat side by side with claims based on emotional, religious and cultural affiliation. Maps, in particular, were put to work here in a variety of ways: to illustrate the importance of taking a standard unit in calculating majority, to show the spread of non-Muslims across the central areas of the Punjab, whether they were in a majority or not, to depict the unity of colonial infrastructure in the form of railways and canals, to predict lines of defence and security, to illustrate the intractable problem of a territorial partition at the smallest of scales, where the border would never be able to create the ideal of homogeneity.

Chapter 5 introduced some of these more marginal maps, produced by Oskar Spate, and explored the less prominent ways that he introduced geographical concepts and ways of thinking into the Muslim League delegation's claim. Spate's observations, experiences and materials do not provide a straightforward narrative of Partition through the eyes of an expert witness. Rather, they work to demonstrate the multiple strands of paradoxical and contentious entanglements that geographers had with the territorial reorganisation processes which marked the transition from empire to independence in India. His work shows how geographical concepts that Mountbatten, the nationalist leaders, and the boundary commissions took for granted were in fact problems for geographers to puzzle over. Contiguity, for example, shifts in tandem with scale. How we see contiguity depends on how we choose to make the map, and following scientific principles is only one of many ways to draw a map. In a process as fraught as a territorial partition, in which affective homelands, property, natural resources, infrastructure, agricultural systems, sites of religious significance and pilgrimage, urban centres and more are at stake, geography's contingencies are understandably exploited.

What was particularly fascinating was that Spate insisted that the Muslim League's claim in Punjab was modest to the point of self-sabotage, while the Congress-Sikh claim overdid it to such a degree that everyone involved knew it. What Spate's work shows is that geography, as a mode of knowing and seeing the world, is as historically constructed and contingent as those modes of statecraft that historians have pointed to when they consider Partition's catastrophic failures. Population statistics, demography and categories of identity are all well-identified as sources of the problem. Geography as an academic discipline and technical practice was similarly limited. Ultimately, there was only so much Spate could do for his employers, and in the end, he could not achieve what he had been hired to do: Qadian went to India and the Ahmadiyya community left for Pakistan, where they have been subject to exclusion and marginalisation by the state because they are legally deemed to be 'non-Muslim', the very category the boundary commissions used to draw the boundaries (Ahmed, 2010).

The Problem with 'Expertise'

The emphasis in Chapter 2 on the role of geography in producing a territorial order in the service of empire somewhat obscures the more complex and sometimes ambivalent ways that individuals may have participated in colonial projects. This speaks to wider contemporary debates about the role of individual scholars when they participate in work beyond the academy. What are their responsibilities to their employer or sponsor? What are their responsibilities to other stakeholders or to publics who may be affected by their work? What if their findings conflict with the needs or aims of the project? How do the limits of their discipline impact their ability to contribute? Oskar Spate's role in the boundary commission illustrates a contentious relationship between the expert and the empire in the 20th century. Spate was employed not by the government but by an interested party; as such, he was not a representative of the state but rather a technical advisor. Interestingly, while it was not uncommon for academics to be broadly acquiescent to empire, Spate was in many ways critical of the state and generally in favour of decolonisation and independence. He wrote in his memoirs that, during his time as an academic at the University of Rangoon before the start of the Second World War, he found himself on the side of his students who were agitating for independence (Spate, 1991).

Spate's training and expertise, however, were still framed by broader geographical modes of knowledge production, and as such, were complicit in and informed by the imperial world order. He is, therefore, an interesting and highly relevant figure, both in his own time and in ours. How effective can an 'expert' be if they are required at all times to remain neutral and objective while working in an applied or prescriptive capacity? How effective could Spate's imperially informed geographical training be in resolving the territorial and social debates which underpinned the partitioning process? Was a British geographer, no matter how progressive, equipped with the discursive or technical tools to significantly alter or influence the final outcome? Spate himself believed that he had made very little difference in the Punjab, and he was deeply saddened by the outcome. He recognised that geographical expertise was largely ignored by the government, perhaps in part because, as he put it, 'if geographers were kings, there would have been no Partition' (Spate, 1947, p. 214).

Timothy Mitchell (2002) writes about the role of expertise and what he calls 'techno-politics' in the construction of the modern state. Mitchell calls attention to the ways in which expertise, particularly scientific expertise, was (and is) put to work for the purpose of constructing the physical infrastructure and built environment of the state. For Mitchell, the politics of the state works to appropriate and mobilise the language and science of expertise in order to claim the knowledge and authority of expertise for itself. Expertise gets drafted, so to speak, for the purpose of consolidating state power and control. In a sense, the Punjab Boundary Commission worked in such a way, whereby 'politics' (in the form of institutionalised political parties and high-level negotiations and commissions) mobilised the rhetorical and visual norms of cartographic practice while bypassing, for the most part, the potential contributions of an objective or neutral expert. Spate's own expertise was pulled by multiple competing forces, as he found himself contracted not simply by the Muslim League but also by the transitioning state's conflicting imperatives to both protect lives at the popular level and to negotiate political compromise at the elite level.

The book's historical work on Oskar Spate's contributions to the Muslim League's presentations to the Punjab Boundary Commission also speaks to the role and limits of geographical expertise and authority outside the academy. Contemporary scholars often find themselves acting as expert consultants in a variety of nonacademic contexts, and Spate's position is not necessarily so far removed from more contemporary contexts. How can and should scholars negotiate their own politics, the politics of the stakeholders with whom they are working, and the politics underpinning the project they may be contributing to? Is neutrality or objectivity either possible or desirable?

Bridging the Gap: Historical Geographies of Partition

The book is deeply indebted to the field of Partition studies, inspired by groundbreaking interventions in the historiography of Partition by Ayesha Jalal (1994), Gyanendra Pandey (2001), Yasmin Khan (2007) and Joya Chatterji (1994) among many others. The book makes its own original contributions to this field in two key ways. First, in building on recent interventions by scholars working on spatialising the historiography of Partition (e.g. Ansari and Gould, 2019; Gould and Legg, 2019), the book demonstrates how historical geographical modes of enquiry and analyses open up new ways of examining Partition and of conceptualising the Partition archive. The colonial geographical archive introduced and analysed in Chapter 2 sheds new light on how colonial forms of geographical surveying and mapping remained consistent from the late 19th century into the 20th century, and how they produced certain norms in the cartographic representation of the subcontinent and its populations. Second, the book unsettles the dominant trend in Partition studies to consider geography, as both academic or professional discipline and technical practice, as a relatively fixed and stable concept in contrast to the socially constructed and politically contingent nature of other forms of colonial knowledge (most especially around cultural and social identity and identification). While scholars like Lucy Chester (2009) and Sarah Ansari and William Gould (2019) have highlighted the importance of locating and placing the political and legal processes of Partition, this book goes further to demonstrate how geographical knowledge and geographical techniques of bordering and boundary-making were themselves contingent and power-laden, and very often embedded in the British colonial project in India.

The historical geographical study of Partition undertaken in the book shows in detail how we might mobilise the history of what Felix Driver (1992) called 'geography's empire' in British India to understand how colonial geography was put to work, both explicitly and in a more subtle fashion, for the purposes of decolonisation and territorial partition. I challenge the notion that more, better or different geographical expertise or data could have substantially changed the outcome (beyond, of course, the geographical expertise that might have cautioned against a partition at all, which Oskar Spate did (1943)). Spatial concepts that sat at the heart of the partitioning process, such as territory, contiguity, political or administrative units, even the idea of a 'workable boundary', are as contingent and historically constructed as the social categories and political ideas that scholars widely and correctly recognise as such. Similarly, geographers themselves were working within the context of their discipline, which had colonial roots. This was particularly true for white European geographers but as Sumathi Ramaswamy (2017) has shown, European geography had long been incorporated into Indian education. And as I have shown in my analysis of the geographical data used in some of the territorial claims put forward by the Muslim League and the Congress Party to the Punjab Boundary Commission, debates about the accuracy of the data itself were common, but challenges to the cartographic norms and geographical *principles* underpinning contiguity and majority were not. The nature and value of geographical data were assumed rather than probed, both in 1947 and in the scholarship of Partition that came later.

The book makes a significant empirical and methodological intervention in the critical history of cartography tradition. Inspired by Matthew Edney's (2019) provocative argument that there is no such thing as cartography and that scholars should instead be examining the ideal of cartography, the book identified evidence of a colonial ideal of cartography at work in British India. The cartographic analysis presented in the book moves beyond a critical analysis of the maps themselves to show how the geographies of Partition were rooted in part in a colonial ideal of cartography. In examining an eclectic mix of maps, surveys, court transcript hearings, speeches, political pamphlets, diaries and more, I have moved beyond a critical cartographic analysis of maps to show how the colonial ideal of cartography could never be fully achieved in British India, and neither could the impossible political ambitions of a territorial partition.

Throughout the book, I conduct critical cartographic analyses of maps and other forms of geographical data, which opens up the possibility of considering how Partition was one eventual outcome of a longer history of mapping and boundary-making that produced British India in certain ways. This enables us to see how Partition came to be articulated as a possible territorial solution to a social and political conundrum facing the British colonial government and the Indian elite during the final years of the Raj. Partition itself, on the basis of religious or communal difference, was not imagined in any clear detail until the 1930s, and cannot be read back into the 19th century. However, cartographic representations of the subcontinent's regions, populations and environments were increasingly standardised in the 19th and early 20th centuries. This demonstrates that though Partition may have been relatively sudden, and the process itself certainly was rushed, the foundational cartographic gaze required for imagining Partition had already been at work in British India for a number of decades.

While the book makes use of the literature produced by historical geographers of empire and decolonisation to introduce new ways of considering geographical knowledge within Partition studies, I also draw attention to the relative paucity of historical geographical writing on Partition. Given the richness and depth of the field of Partition studies, as well as its inherent interdisciplinarity, with contributions from scholars whose work spans the humanities and social sciences, this gap in historical geography is fascinating. It perhaps reflects a wider trend in the subdiscipline that has tended to overlook the period of decolonisation in the mid-20th century.

The book's extended timeline, beginning in the second half of the 19th century and culminating in the summer of 1947, enables me to make a dual, and therefore unique, contribution to both the more established debates on the historical geographies of empire and to the emerging debates on the historical geographies of decolonisation. The four chapters working in tandem tell a story about colonial geography as it made its way from disparate survey projects into standardised and accessible formats and finally into new nationalist visions of India that challenged the colonial project it originally enabled. Colonial geographies of India are long-lasting. In Chapter 2, I showed how the *Imperial Gazetteer of India*, a relatively under-examined colonial archive (especially compared to the better-studied and more thoroughly theorised census), was in fact a significant attempt to put geographical data to work in the service of colonial administration. The gazetteers represent an untapped resource for more thoroughly understanding the ways that colonial India was rendered geographically, and for understanding how colonial geographical sources themselves narrated the historical geography of India at the time. However, the book also makes another key contribution to debates in historical geographies of empire. Inspired by accounts of Partition that have grappled with the ways in which Partition involved both British and Indian actors, I draw attention to the fact that much of the scholarship by historical geographers focuses on colonial geographical practice, colonial maps, and often white European or American individuals.

The book therefore makes use of one of the key postcolonial contributions of the Partition studies literature: that Indians were not only responding to, appropriating or rejecting the colonial geography outlined in Chapter 2, but they were conceptualising and imagining India's historical, present and future geographies in alternative ways, even if they were not always articulating their visions in the narrow rhetoric of so-called 'British geography'. While historians of Partition have long understood the dynamic interplay between European colonial knowledge and indigenous and colonised elite forms of knowledge, historical geographies of empire have until recently focused more on colonial knowledge and the ways in which it attempts to dominate, erase or appropriate local knowledge and culture. The book therefore introduces and reinterprets through a geographical lens a selection of materials produced by Indian Muslim thinkers and nationalist leaders from the late 19th and early 20th centuries. Some of these materials are well examined, including Sayyid Ahmad Khan's two-nation theory, Muhammad Ali Jinnah's Lahore Resolution of 1940, and the later maps of Dinia published in pamphlets by Choudhary Rahmat Ali. However, here they have been subjected to an original analysis which has highlighted the distinctive geographical imaginations that they produce. As importantly, here they have been brought into conversation with the knowledge produced by colonisers, demonstrating the connections between these visions, often analysed discretely.

The book also speaks to the emerging debates in the historical geographies of decolonisation, examining in detail both the spaces of the partitioning process in 1947 and the ways in which the partitioning process imagined the political, social and sacred spaces of India. The Punjab Boundary Commission, which heard arguments in the Lahore court over 10 days in July 1947 before Radcliffe drew his final boundary in Simla, was one of a number of organisations that facilitated India and Pakistan's passage from British colony to independent states. The commission, and especially Radcliffe, was undertaking fundamentally geographical work, whether they framed it as such, and so these chapters highlight how geography was both central to and sidelined by the political and legal concerns surrounding decolonisation in British India. Geographers themselves occupied less prominent, and certainly less powerful, positions throughout the process of Partition, sitting to one side of the decision-makers, never making it into the room where the final boundaries were drawn. Yet the geographical techniques of boundary-making and map-making underpinned the politics as they played out, a pattern that scholars might see play out in other historical decolonising projects in other regional contexts. More empirical work needs to be done, however, in order to understand how geographers involved themselves (or not) in decolonisation processes around the world in the mid-20th century. Spate (1991) later observed that very few British geographers had concerned themselves overly much with what was happening in India in the 1940s. Were they overlooking Britain's retreat from her empire more generally, focusing instead on rebuilding Britain after a devastating global conflict? As Dan Clayton has recently written, 'Is it possible to talk of the passing of geography's empire after 1945? If so, how did this passing manifest itself to geographers. . .? Did Western geographers recognize their complicity in empire. . .?' (2020, p. 1543).

Taken together, the four empirical chapters of the book showed how older colonial geographies were put to work in new ways for dismantling the empire. Decolonisation relied in large part on partial, problematic data that had often been reproduced or reinterpreted, and had often been severed from the contexts and contingencies of its creation. Colonial boundaries and territorialised identities were first mobilised and then reinscribed by the partitioning process, illuminating how colonial geographical knowledge was embedded in the geographies of the postcolonial independent states of India and Pakistan from the moment the new borders were drawn.

Finally, the conclusions drawn in the book about the shifts in real-world applications of geographical knowledge, particularly by politicians and bureaucrats in the United States and Europe for the purposes of boundary-making and reordering territory, demonstrate the underlying ideology of standardisation and universalisation at work in geographical thinking in the early to mid-20th century. This has profound consequences for scholarship on partitions, and on borders and boundaries more broadly. Despite the book's empirical focus on a single partition context, the book nevertheless points to the ways in which geographical techniques and methods were being systematised and implemented in multiple and diverse contexts in the 20th century. Territorial partition was a significant aspect of 20th-century decolonisation processes, as the British retreated from their empire, not only in India but in Ireland and Palestine as well (not to mention the French withdrawal from Indochina).

The Partition of India and Pakistan should therefore be understood as one iteration of a more systematic *geographical technique* that has been used in disparate contexts to manage difficult territorial conundrums. The territorial partition, despite its many acts of violence in former colonies (Palestine, Ireland, Vietnam) and Cold War contexts (Germany, Korea, Vietnam), continues to crop up in high-level political discourse. For a brief period, during the writing of this book, partition was floated as a potential solution to Russia's invasion of Ukraine in the winter of 2022. There is an underlying geographical logic to the concept of a partition that can be understood in terms of the theoretical frameworks used here: the cartographic ideal, and how it informs the deployment of maps and mapmaking as tools of governmentality. Partition in technical geography terms, then, is a *method*, a *technique*. That method, therefore, has its own genealogy, and that genealogy is still waiting for a historical geographer to write it.

References

Ahmed, A.A. (2010). The Paradoxes of Ahmadiyya Identity: Legal Appropriation of Muslim-ness and the Construction of Ahmadiyya Difference. In *Beyond Crisis: Re-evaluating Pakistan*, N. Khan, 273–314. New Delhi: Routledge India.

Ansari, S. and Gould, W. (2019). *Boundaries of Belonging: Localities, Citizenship and Rights in India and Pakistan*. Cambridge: Cambridge University Press.

Chatterji, J. (1994). *Bengal Divided: Hindu Communalism and Partition, 1932–1947*. Cambridge: Cambridge University Press.

Chester, L. (2009). *Borders and Conflict in South Asia: the Radcliffe Boundary Commission and the Partition of Punjab*. Manchester: Manchester University Press.

Clayton, D.W. (2020). The Passing of "Geography's Empire" and Question of Geography in Decolonization, 1945–1980. *Annals of the American Association of Geographers* 110, 1540–1558. https://doi.org/10.1080/24694452.2020.1715194.

Driver, F. (1992). Geography's Empire: Histories of Geographical Knowledge. *Environment and Planning D: Society and Space* 10, 23–40. https://doi.org/10.1068/d100023.

Edney, M. (2019). *Cartography: The Ideal and Its History*. Chicago: University of Chicago Press.

Gould, W. and Legg, S. (2019). Spaces Before Partition: An Introduction. *South Asia: Journal of South Asian Studies* 42, 69–79. https://doi.org/10.1080/00856401.2019.1554489.

Jalal, A. (1994). *The Sole Spokesman: Jinnah, the Muslim League and the Demand for Pakistan*. Cambridge: Cambridge University Press.

Khan, Y. (2007). *The Great Partition: The Making of India and Pakistan*. New Haven; London: Yale University Press.

Mitchell, T. (2002). *Rule of Experts: Egypt, Techno-Politics, Modernity*. Berkeley: University of California Press.

Pandey, G. (2001). *Remembering Partition: Violence, Nationalism and History in India*. Cambridge: Cambridge University Press.

Ramaswamy, S. (2017). *Terrestrial Lessons: The Conquest of the World as Globe*. Chicago: University of Chicago Press.

Scott, D. (1995). Colonial Governmentality. *Social Text* 191–220. https://doi.org/10.2307/466631.

Spate, O.H.K. (1943). Geographical Aspects of the Pakistan Scheme. *The Geographical Journal* 102, 125–136. https://doi.org/10.2307/1788890.

Spate, O.H.K. (1947). The Partition of the Punjab and of Bengal. *The Geographical Journal* 110, 201–218. https://doi.org/10.2307/1789950.

Spate, O.H.K. (1991). *On the Margins of History: From the Punjab to Fiji*. Canberra: National Centre for Development Studies, Research School of Pacific Studies, Australian National University.

Index

academics, Western, 140, 141, 184. *see also* objectivity, tension with in applied work
accuracy
  in boundary-making, 67, 70, 121
  of census data, 22, 152, 164, 170
  in colonial knowledge production, 49, 93–94
  of delegation maps, 171–173
  in geographical knowledge, 70, 114–115, 120
Ahmad, Asad, 158–159
Ahmad, Kazi S., 149
Ahmadiyya
  delegation of, 141–142, 144, 157–158
  as religious minority in Pakistan, 158–159
Ahmadiyya movement, 84–85
Ahmad Khan, Sayyid, 78, 85, 87–88, 91–94
All-India Muslim League. *see* Muslim League
Amritsar, 134–135, 149
Amritsar district, 118, 154, 159–163
Anderson, Benedict, 102
Ansari, Sarah, 4, 13, 83
anti-colonialism, 7, 22–23, 85, 94, 140
Asif, Manan Ahmed, 87–89
authority. *see also* scientific discourse
  circular, 49, 55, 56
  of gazetteers, 48–49, 51
  geographic, 46, 67, 119, 143, 168
Azad, Maulana Abul Kalam, 11, 82, 94

Baluchistan, 97, 101, 125
Barelwi, Sayyid Ahmad, 84, 85
Bengal. *see also* violence, communal
  academic histories of, 10, 13
  maps of, 40, 43, 125
  proposal for independent, 101, 160
  as resource-rich, 121
  shared language in, 102
Bengal Boundary Commission. *see* boundary commissions, in India
Bengalis, 66, 86, 92
Bharat Mata, 22, 89, 98
Bikaner, 117
border disputes, 116
borders, 6, 66–67, 69–72, 90
boundaries, 125, 168–170
boundary commissions, general, 121

Mapping Partition: Politics, Territory and the End of Empire in India and Pakistan, First Edition. Hannah Fitzpatrick.

boundary commissions, in India
hearings of, 9, 26, 118
process of, 116–119, 127
structure of, 8–9, 117
boundary-making
as art and science, 65, 68–73, 176
avoidance of migration in, 102
as colonial practice, 41, 62, 65–67, 103, 141
high stakes of, 117, 176
importance of defence in, 167
as inherently geographical, 116, 188
linked to social and political boundaries, 103
scientific discourse of, 66–67
standardisation of, 64
Boundary-Making: A Handbook for Statesmen, 68, 69, 73
Bowman, Isaiah, 68, 69
buffer regions, 107, 163, 168

Cabinet Mission Plan, 8, 108
Calcutta, 9, 117, 118, 159
canals, 38, 54, 122–124, 135, 161. *see also* infrastructure
agricultural development and, 38, 122, 161–163
in Partition claims, 133, 135, 150, 157, 161
cartography. *see also* Imperial Gazetteer of India Atlas
colonial ideal of, 21, 23, 186–187
by colonised people, 22–33, 103–104
critical history of, 20–24, 68
in decolonisation, 125
ethnographic mapping in, 68–69, 121, 125
in Imperial Gazeteer of India Atlas, 124–125
limits of population mapping in, 173
obscuring diversity, 58, 135
providing legitimacy, 121, 177
territorial control and, 41, 51, 125
caste
category of, 46, 54, 66, 104, 127
as electoral group, 7, 91, 92, 127
census
cited as unreliable evidence, 129, 134, 146, 164, 170
as data for delegations, 124, 146, 149
first all-India, 46
historiography of, 38–39
power of, 39, 146
and production of modern identities, 22
Chatterjee, Partha, 100
Chatterji, Joya, 116, 120, 143
Chester, Lucy
on boundary commissions, 14, 17, 117
on colonial maps, 67
on logic of Indian responsibility, 116, 118, 120
Chishti Sufism, 82–83
Clayton, Dan, 18, 20, 189
colonial rule
justification of, 47, 50, 55, 62
as progressive, 19, 37, 44, 45, 61
communalism, 96, 99–100, 145. *see also* violence, communal
Congress Party (INC)
arguments to boundary commissions, 129, 133–135, 164, 168, 170
Cabinet Mission Plan and, 108
delegation, 117, 122, 164
discourse of Indian responsibility and, 120
dissenting voices in, 94
lack of consideration for diversity, 92
motivation to keep resource-rich areas, 121
move towards Hindu-inflected secularism, 99
narratives around Partition and, 100
nationalism and, 7
in nomination of judges, 9, 115
politics of, 7, 94, 97, 98
rhetoric of unity over diversity, 92, 135
use of 'other factors,' 133–134, 164
Congress red map
gerrymandering in, 129, 132, 150, 170–171
unity in, 134–135

Congress-Sikh claim, 150, 164, 167–168, 175
Congress-Sikh line, 119, 136, 164–165, 168
contiguity, territorial, 121, 128, 132–133, 149–150
corridors, proposed, 124, 159–161
courtroom, the, 26–27, 118–119, 141

Dalit (scheduled castes), 7, 117
data, geographical
  collection methods, 47, 49
  collection sources, 54, 57
  colonial ideals surrounding, 45
  as evidence, 118, 119
  inconsistency of, 50, 62, 67, 129
  standardisation of, 44–46, 129, 189
  subordinate to legal and political concerns, 116
  in territorial conquest, 21
decolonisation, 20, 23, 109–110, 156, 188–189. *see also* postcolonialism
defence, 148, 150, 164–167. *see also* buffer regions
Delhi, 118, 124, 163, 167
delimitation, of boundaries, 70, 71
demarcation, of boundaries, 70, 71
democracy, electoral
  census categories and, 39, 97–98
  minorities and, 7, 99, 103, 121, 132
  territorial divisions and, 97
Deoband school, 84
Devji, Faisal, 78, 93, 95, 98, 156
Dinia, 103–105, 107, 129
Driver, Felix, 20, 21
Durand Line, 116

East India Company, 38–40, 43, 45, 87
Edinburgh, home of cartographer Bartholomew, 125–126
Edney, Matthew, 20–23
education reform, Muslim, 84–86
Elden, Stuart, 149, 152
elections of 1945, 8
enclaves, proposed, 159–161
English East India Company (EEIC). *see* East India Company
environmental determinism, 66
ethnographic mapping. *see* maps, ethnographic
expertise, geographical, 5, 120, 140, 177, 185

Feudatory states, 50, 105
Foucault, Michel, 21–22, 41
Fraser, Thomas, 14
frontiers, 62–65, 146. *see also* North-West Frontier Province

Gazetteer of the Territories under the Government of the East-India Company and the Native States on the Continent of India, 44–45
Gazetteer of the Territories under the Government of the Viceroy of India, 48–49
gazetteers. *see also* individual gazetteers
  context of production, 42–43
  early history of, 38–46
  population diversity in, 55, 61
  scope and scales of, 43, 50, 54
  as spatial imaginary of colonial control, 41–42
  systematic data collection in, 46–47
  territorial history and, 51–54
  as untapped resource, 187
geographers, 140–141, 188–189
Geographical, Statistical, and Historical Description of Hindostand and the Adjacent Countries, 43
geographical discourse, marginalised by legal process, 150, 152
geographical knowledge. *see* expertise, geographical; knowledge, geographical
geographical practices, relative ontological stability of, 23, 79–80
geography. *see* decolonisation; historical geographies; postcolonial geography
gerrymandering, 129, 132, 150, 170–171, 177

Ghulam Ahmad, Mirza, 84–85, 156
Gilmartin, David, 13, 15, 16, 38, 83
globes, 86
Goswami, Manu, 86, 89, 97–98
Gould, William, 4, 13, 79, 82
governmentality, colonial, 21–22, 39, 44, 48, 57, 91
Government of India Act, 35, 45, 98
Great Trigonometrical Survey (GTS), 21, 40
Gurdaspur, 118, 134–135, 150, 154, 156

Hamza, El, 145
Harley, J.B. (John Brian), 20–21, 23
Hindu population
  in Congress claim, 134, 136
  homeland of, 87–88
  in maps, 51, 125
  in politics, 7, 39, 92, 95
  in statistics, 95, 96, 103, 105, 170
historical geographies
  colonial, 88–89
  of decolonization, 4, 5
  of empire, 5, 17–18
  indigenous knowledge and, 177
  perspectives of, 185–189
Historiographies of Indian Partition
  avoidance of by religious studies, 15
  critical perspectives on, 11–13
  early accounts of, 10–11
  on identity and violence, 4
  lack of geographical perspective in, 15, 17
  in political geography, 14–15
  variety of narratives in, 10, 13
Holdich, Thomas
  and buffer regions, 107
  on dangers in Partition process, 117
  on delimitation *vs* demarcation, 71–72
  on frontiers, 63–67, 168
  on local involvement, 116–117
  on natural boundaries, 168
homeland for Hindus, 87, 88
homeland for Muslims, 101, 107–108, 146
homeland for Sikhs, 173
homelands, 88, 103–105, 108. *see also* homeland by population category
homogeneity and political citizenship, 91, 92, 96, 98, 134
Hunter, William Wilson
  background of, 45, 47
  Imperial Gazetteer of India and, 35, 47
  'Indian history' and, 50–55
  methods of, 48–49, 54–55

identities, diversity of
  captured *vs* produced, 22, 54
  census categories and, 39, 42
  in colonial knowledge production, 54, 55, 87, 88, 107, 125
  conflation of political and religious, 16, 99
  ethnic, 22, 47–48, 54
  fixed onto territorial units, 51, 54, 58, 182
  national unity and, 96–97
  overlapping categories of, 54, 96–97
  racial, 22, 47–48, 54
  religious, 54, 56–58, 87–88, 90–91, 96
  role in boundary-making, 61, 64, 121
  as source of conflict, 92, 99
imaginaries, cartographic and territorial
  of anti-colonialism and nationalism, 74, 79, 80, 87, 94, 102
  of colonial control, 40–42
  examination of by geographers, 19
  homogeneity of, 155
  territorial othering of Muslim population and, 89
Imagined Communities, 102–103
Imperial Gazetteer of India
  format of, 55
  plan of, 46–48
  population diversity and, 55–58, 61
  as used by Holdich, 64–65
Imperial Gazetteer of India Atlas, 124–125, 127, 145
independence movement, 7–8, 36
independence of Pakistan, 1, 11, 175
Indian Empire, The, 50, 55–57, 62, 66

Indian intellectuals, 39, 78–80, 86, 88–89, 98
Indian National Congress (INC). *see* Congress Party
indigenous forms of knowledge, 23, 177, 188
infrastructure
  in boundary-making claims, 135, 150, 164–167
  importance of, 122–124, 161–162, 170
Iqbal, Muhammad, 78, 82, 89–90, 96–97

Jalal, Ayesha, 11, 14, 15, 95
Jazeel, Tariq, 19–20
Jinnah, Muhammad Ali
  political strategies of, 8, 11, 101
    conceptions of secular Pakistan, 96
    and two-nation theory, 78, 90
  on religious *vs* political identities, 98–99
Jones, Kenneth, 39, 57, 80, 81, 146
Jones, Reece, 14–15
Jones, Stephen B., 68–73, 116

Kashmir, 101, 134–135, 150
Khan, Muhammad Zafrullah
  arguing for standard units, 129, 132
  arguing for workable boundary, 121
  as delegation representative, 117, 144
  on strategic maps, 165, 167
Khan, Yasmin, 6–8, 13, 116, 118
knowledge, colonial
  administration and, 43–44, 66, 67, 182
  reformers and nationalists and, 106–107
knowledge, geographical
  authority of, 143, 177
  empire and, 18–20, 140–141, 183
knowledge production, colonial. *see also* census; gazetteers; knowledge, colonial; knowledge, geographical
  change in mode of, 50
  and development in the Punjab, 38
  differing contexts of, 55, 56
  and the East India Company, 43
  lack of objectivity of, 115
  neglected in historiographies, 16

Lahore, 37, 118, 135, 147, 149, 170
Lahore Declaration, 107–108
Lahore Resolution, 125
languages, 95, 101–103, 160
legal discourse, dominance of, 150, 152
legitimisation of British rule, 44, 51, 55, 61
liberalism and Indian intellectuals, 80, 91–92
local involvement, colonial logic of, 47, 116

majority. *see* population, majority
Mapping an Empire, 21
maps. *see also* cartography
  authority of, 51, 74, 127
  Bharat Mata, 22, 89, 98
  in boundary commission testimony, 128–137, 145, 149
  of Dinia, 105
  ethnographic, 68, 121, 125, 127, 136
  homogenising, 136
  in *Imperial Gazetteer of India,* 51
  interpretation of mediated by time, 127–128
  Mughal *vs* British, 48, 104–105
  representing power relations, 22
  strategic, 164–167
martial races, 38, 51, 66
memory, 12, 14
Metcalf, Barbara, 82, 84
migration
  as academic theme, 13
  complicating mapping, 38
  not forewarned by Partition maps, 136–137
  in Partition, 2, 9, 102
  threat of forced, 134
  in two-nation theory, 87
minority. *see* population, minority
missionaries, 39, 81–82
mobility, 82, 87, 89, 124, 134. *see also* migration
Mohammedan Anglo-Oriental College (MAO), 85
Mother India. *see* Bharat Mata

Mountbatten, Louis
his narrative of Partition, 11, 176
in Partition process, 8, 108, 115, 116, 120, 152
Mughal Empire. *see also* maps, Mughal *vs* British
in colonial histories, 57–58, 88–89
Muslim scholarship on, 89
Punjab history and, 37, 84
religious leaders and, 82
Muslim exceptionalism, 82, 88, 96, 102
Muslim League
arguments to boundary commission, 128–129, 133, 145–146, 155, 165
Cabinet Mission Plan and, 108
delegation, 117, 122, 124, 144
discourse of Indian responsibility and, 120
Muhammed Iqbal address to, 90, 96
in nomination of judges, 9, 115
Partition narratives of, 100
politics of, 8, 95, 97, 98, 136
Muslim League claim, 128–133, 145, 155, 164–165, 167
Muslim League line, 119
Muslim League map, 129, 133, 134, 136–137, 165
Muslim population
as historically constructed and politically contingent, 95
as majority population, 103, 125
as minority, 94–96
Muslim reform movements, 82–85
mutiny, 51, 54
Mysore gazetteer, 48

narratives, colonial
embedded in nationalist thinking, 100
of migration *vs* stasis, 80, 89
of race and religion, 57, 127
territorial history and, 50–51
totalising, 44, 47, 56–57, 97–98
totalising *vs* messy reality, 22, 49, 62
of unity, 89
narratives of new nation-states, 109, 152
national identity, creation of, 77–78, 93
nationalism
colonial pedagogy and, 86
communalism and, 93, 99–100
electoral politics and, 98–99
territorial history and, 51, 56, 98, 102
nationalist movements, 7, 92, 94
developed in context of the colonial state, 97–98, 102–103
as source of conflict, 72, 136
nation-states
cartography claims to, 22
European conceptions of, 80–81, 90–91
as model of sovereignty, 141
post-empire, 20
territorial contiguity and, 121
native customs in colonial administration, 65–66
neutrality, 9, 70, 115. *see also* objectivity
New Delhi, 118
North-West Frontier Province, 43, 97, 98, 101, 125
notional boundary, 118, 120

objectivity. *see also* neutrality
in gazetteers, claims of, 45, 49
in geographical knowledge, 115, 139–140
as goal in boundary-making, 70, 143
tension with in applied work, 143, 147–148, 176–177, 184–185
oral histories, 12–14
order, colonial, 48, 54
through colonial administration, 62
use of boundaries for, 65
Orient, the, 18, 54
Osmanistan, 105
'other factors,' 120–122, 124, 133, 153, 174–175. *see also* Congress Party (INC), use of 'other factors'; Sikh claim
othering, 18, 19, 42, 89

pāk, 102
Pakistan
as acronym, 101

imagined as secular federalised nation-state, 96
as lexeme for population unmixing, 102
as possible bargaining chip, 11, 101, 108
as term embedded in historiographies, 108
Pakistan Declaration. *see* Lahore Declaration
Pandey, Gyanendra, 12, 15, 39, 56, 99
partition. *see also* Partition process, in India; territorial partition
geographical perspectives on, 14–15, 185–189
as geographical term, 16
as term embedded in historiographies, 108
Partition, historical geographical perspectives on, 185–189. *see also* historiographies of Indian Partition
Partition historiographies of India. *see* historiographies of Indian Partition
Partition process, in India. *see also* boundary commissions, in India
administrative structure, 8–9
as both geographical *and* political, 3, 127, 154
context of, 6–7
postcolonialism and, 19–20
Pathankot, 124, 133, 135, 157
pedagogy, colonial, 86
pirs, 83–84
Political Frontiers and Boundary Making, 64
political geography, 14–15, 71, 143, 145, 177
population, majority, 129, 155–156
population, minority
in representative government, 92, 94, 96
as spatial category, 95, 132
population maps in territorial claims, 145
postcolonial geography, 18–20
postcolonialism, 18
postcolonial studies, 100, 140, 188
princely states, 48, 50, 105, 117, 152–153. *see also* Feudatory states
Prophet Muhammad, 84–85
proportional representation. *see* democracy, electoral
Punjab. *see also* infrastructure
colonial histories of, 35–38
geographies of, 36–38, 122
Iqbal's vision of, 97
strategic value of, 107, 121, 163
as a unit in PAKSTAN, 101
Punjab Boundary Commission. *see* boundary commissions, in India
Punjab gazetteer, 54

Qadian, 142, 144, 155–158
Qasmi, Ali Usman, 84–85

race. *see* identities, racial
Radcliffe, Cyril, 9, 14, 115, 118, 122
Radcliffe Award, the, 9, 128, 143
Radcliffe line, the, 9, 120
Rahmat Ali, Choudhary, 94, 101–107, 125, 145, 160
railways, 54, 161, 164, 169, 183. *see also* infrastructure
Ramaswamy, Sumathi, 22–23, 74, 78, 86, 89, 104
rationality, 175. *see also* objectivity
religion and colonial governmentality, 57
religious sites in territorial claims, 134, 158, 173
Renan, Ernest, 96
resources, 121, 124, 153
responsibility, Indian, 26, 116, 120
rootedness, 87, 89, 174

Said, Edward, 18
scales
in anti-colonialism, 94
in census, 39
in deciding majority/minority populations, 129, 132
in gazetteers, 50, 54
in governmentality, 21–22
in Partition historiographies, 13, 14
as problem for representative government, 92
scheduled castes (Dalits), 7, 117
scientific discourse, 56, 67

secularism, 7, 42, 96, 98–99, 136
security. *see also* defence
as concern for Muslims, 90, 96, 107
security and stability
colonial governance and, 54
economic, 90, 96, 107, 153
linked to territorial autonomy, 90
separatism, 107, 173
Setalvad, M.C. (Motilal Chimanlal), 117, 129, 135, 164, 168
Sikh claim, 129, 155, 173–176
Sikh delegation, 117, 122, 134
Sikh independence movement, 173
Sikh population
colonial attitudes towards, 38, 66, 82
historical and religious sites of, 134, 158, 173
on map, 125
as minority, 173, 174
in the Punjab, 122, 153
Simla, 175
Sind, 13, 83, 97
social divisions. *see* identities
sovereignty, territorial, 177
Spate, Oskar
as advisor for Ahmadis, 144
as advisor for Muslim League, 5, 128, 137, 144, 147–150, 154
background of, 142
danger, his recognition of, 117, 175
lecture to the Royal Geographical Society, 154, 158, 173, 176–178
maps of, 118, 124, 148, 164–165
'other factors,' his opinion on, 122, 161, 174–175
perspective of, 143–144, 148, 153, 184
pre-1947 work on India, 144–146
rationality and, 148, 175, 178
reflections on subjectivity, 158
spatial metaphors, 16–17
spatial practices as tools of empire, 19
Subaltern Studies, 12, 99, 100, 110
Sufism, 82, 83
Survey of India, 21, 40, 109, 122
surveys, 22, 41, 46, 47. *see also* census; gazetteers; Survey of India
tehsils, 128–129, 132–133, 136
terms of reference of boundary commissions, 120–124, 150, 154
territorial contiguity. *see* contiguity, territorial
territoriality, 141, 154, 158, 177
in relation to nation-states, 141
territorial partition
examples of, 14, 121, 156, 189
logic of, 62
territory. *see also* imaginaries
construction of as political and technical, 149, 152
Thornton, Edward, 44, 48
two-nation theory, 78, 86–91

'ulama, 82–84
umma, 90, 136
Unionist Party, 7
United Provinces, 7, 95, 98
unity, as colonial construct, 40, 107, 134, 145
Urdu language, 95, 106, 145

vagueness
of boundary-making processes, 121
of commission terms of reference, 120, 154, 175
of rootedness claim, 174
of terms used in Congress claims, 135
violence, communal, 8, 9, 99
as concern of Oskar Spate, 148
history of, as part of Sikh claim, 173–174
linked to colonial classification of differences, 62
linked to Partition process, 2, 15, 116, 121
in Partition historiographies, 12, 120
as spur to revisit history of Partition, 12
visibility of good boundaries, 168–170

Weizman, Eyal, 119
workable boundary, 140, 141, 160–161, 165, 168

Yong, Tan Tai, 38, 51, 122